Eurotherm Seminar No 108 – Nanoscale and Microscale Heat Transfer V

Journal of Physics: Conference Series Volume 785

Santorini, Greece
26 – 30 September 2016

Editors:

Konstantinos Termentzidis
Xanthippi Zianni
David Lacroix

ISBN: 978-1-5108-3534-4
ISSN: 1742-6588

Printed from e-media with permission by:

Curran Associates, Inc.
57 Morehouse Lane
Red Hook, NY 12571

Some format issues inherent in the e-media version may also appear in this print version.

Printed by Curran Associates, Inc. (2017)

For permission requests, please contact the Institute of Physics
at the address below.

Institute of Physics
Dirac House, Temple Back
Bristol BS1 6BE UK

Phone: 44 1 17 929 7481
Fax: 44 1 17 920 0979

techtracking@iop.org

Additional copies of this publication are available from:

Curran Associates, Inc.
57 Morehouse Lane
Red Hook, NY 12571 USA
Phone: 845-758-0400
Fax: 845-758-2633
Email: curran@proceedings.com
Web: www.proceedings.com

Table of contents

Volume 785

Eurotherm Seminar No 108 - Nanoscale and Microscale Heat Transfer V
26–30 September 2016, Santorini, Greece

Accepted papers received: 6 December 2016Published online: 16 January 2017

Preface

Papers

IOP Conf. Series: Journal of Physics: Conf. Series **785** (2017) 011001 doi:10.1088/1742-6596/785/1/011001

Nanoscale and Microscale Heat Transfer V (NMHT-V)
EUROTHERM seminar No 108

Konstantinos Termentzidis[1,2]*, Xanthippi Zianni[3] and David Lacroix[1,2]

[1] Université de Lorraine, LEMTA, CNRS, UMR 7563, 2 avenue de la Forêt de Haye, TSA 60604, 54518, Vandœuvre les Nancy cedex, FRANCE

[2] CNRS, LEMTA, UMR 7563, 2 avenue de la Forêt de Haye, TSA 60604, 54518, Vandœuvre les Nancy cedex, FRANCE

[3] Technological Educational Institution of Sterea Ellada, Dept. of Aircraft Technology, 34400 Psachna, GREECE

Electronic address: * Konstantinos.Termentzidis@univ-lorraine.fr

Preface

This special volume of JPCS contains the Proceedings of the **5th Nanoscale and Microscale Heat Transfer (NMHT-V).** The conference was held in Santorini, Greece, from September 26th to September 30th 2016. The NMHT Eurotherm seminar series aim to give the state-of-the-art and to present the most recent advances of the scientific research in the field of heat transfer at nanoscale and microscale where classical macroscopic heat transfer laws might be not applicable. NMHT-V is the 5th of the Eurotherm seminars. Previous seminars were held in Lyon (October 2014), Poitiers (1998 & 2011) and Reims (2003).

NMHT-V had more than 100 participants, among them 30 PhD students, mainly from Europe (England, France, Germany, Spain, Italy, Greece, Ukraine, Finland, Russia…15 countries) but also from countries outside Europe (U.S., Japan etc). Universities and research laboratories doing pioneering work in the scientific field of the seminar were represented.

During the 5 days of the conference (26-30 September 2016), scientists and researchers presented and discussed their most recent advances through more than 75 talks during 17 sessions. Furthermore, a poster session with 30 presentations had allowed fruitful exchanges between. A Special Flash poster presentation session (2 minutes for each poster) was included on the first day of the conference to give the opportunity to the researchers/students to introduce their work to the audience. The scientific program included keynote lectures and tutorials. Moreover, the annual one-day (September 28th) workshop of the European research project "Quantiheat": http://www.quantiheat.eu/ was hosted on the conference location and was open to all participants interested in quantitative measurements at micro/nanoscale and thermal imaging.

The conference organisers gratefully acknowledge the members of the scientific committee and the experts who carried out the reviews of abstracts and of proceeding articles. Also acknowledged is the support by the Société Francaise de Thermique, LEMTA laboratory, CNRS, University. of Lorraine, Quantiheat European Project (S. Gomes), The European Research Group for Heat Transfer at the Nanoscale (S. Volz) and the Institute of Optic (P. Ben-Abdallah).

The next NMTH conference is scheduled in Lapland, Finland, December 2018 and will be organized by Professor Mika PRUNILLA and the VTT (http://www.vttresearch.com)

NMHT-V focused on:

- Conductive heat transfer at short scales of length and time,
- Thermal radiation at subwavelength scale,
- Heat and mass transfer in solid/liquid interfaces

The main topics of the seminar were:

- Phonon, electron, mass transport
- Thermal Metrology at Micro/Nanoscales
- Computational Methods in Micro/Nanoscale Transport
- Micro/Nanoscale Interfacial Transport Phenomena
- Micro/Nanoscale Boiling and Condensation Heat Transfer
- Micro/Nanoscale Thermal Radiation
- Thermoelectricity/Thermophotovoltaics
- Biomedical Applications of Micro/Nanoscale Transport

Chairs

K. Termentzidis,	CNRS researcher,	LEMTA-CNRS, Nancy, France
D. Lacroix,	Professor,	LEMTA-Un. Lorraine, Nancy, France
X. Zianni,	Professor,	TEI of Sterea Ellada, Greece

Scientific Committee (22 members)

J-H. Bahk,	Un. Cincinnati, U.S.
P. Ben-Abdallah,	Institut d'Optique, Orsay, France
O. Bourgeois,	Institut Néel, Grenoble, France
Y. Chalopin,	EM2C, Châtenay-Malabry, France
P.-O. Chapuis,	CETHIL, INSA-Lyon, France
S. Dilhaire,	LOMA, Un. Bordeaux, France
D. Donadio,	Max-Planck, Germany
S. Gomes,	CETHIL, INSA-Lyon, France
A. Kent,	Un. Nottingham, U.K.
A. Kittel,	Un. Oldenburg, Germany
E. Lampin,	IEMN, Un. Lille 1, France
V. Lysenko,	INL, INSA-Lyon, France
J. Lukes,	University of Pennsylvania, U.S.A.
M. Martin-Gonzalez,	IMM, Spain
S. Merabia,	Intitut Lumière Matière, Lyon 1, France
N. Papanikolaou,	Demokritos, Greece
G. Pernot,	LEMTA, Un. Lorraine, France
B. Perrin,	INSP – Univ. Paris 6, France
S. Shevlin,	Materials Chemistry, UCL, U.K

C. Sotomayor, ICN2, Spain
R. Vaillon, CETHIL, INSA-Lyon, France
S. Volz, EM2C, Châtenay-Malabry, France

Tutorials from the invited speakers:
Six tutorials were given by distinguished scientists aiming to give to young researchers an overview and background knowledge of the scientific topic of the conference.

D. Cahill, Univ. of Illinois at Urbana-Champaign,USA,
"Time domain thermoreflectance, Fundamentals"
K. Joulain, Un. of Poitiers, France,
"Introduction to fluctuationnal electrodynamics"
P. Keblinski, Rensselaer Polytechnic Institute, USA,
"Determination of Thermal Conductivity and Interfacial Thermal Resistance by Molecular Dynamics Simulations"
D. Poulikakos, ETH Zurich, Switzerland,
"How to arrest, interrogate and transport viruses in liquids one at a time"
D. Jaque, Un. Autonoma de Madrid, Spain,
"Luminescence nanothermometry: Fundamentals and Applications"
A. Togo, Un. Kyoto, Japan
"First-principles lattice thermal conductivity calculation and phonon database"

Peer Review Statement

The guest Editor of this volume of Journal of Physics: Conference Series is K. Termentzidis. All papers have been reviewed by expert referees among the Scientific Committee of the NMHT-V to the professional and scientific standards expected of a proceedings journal published by IOP Publishing.

Proceeding articles (alphabetic order)

[1] A. Alkurdi and S. Merabia, "Thermal transmission at Si/Ge interface: ab initio lattice dynamics calculation"
[2] J-L. Battaglia, I. De, V. Sousa, "Inverse heat conduction problem in a phase change memory device"
[3] S. Fumeron, B. Berche, F. Moraesa and F. Santos, "Tayloring energy levels with curvature? An illustration of Da Costa formalism"
[4] E. Hatzikraniotis, Th. Kyratsi, K.M. Paraskevopoulos, "Modeling of thermal conductivity in high performing thermoelectric materials"
[5] T-P. Nguyen, L. Thiery, D. Teyssieux, D. Briand and P. Vairac, "Recent improvements on micro-thermocouple based SThM"
[6] H. Ghasemi and A. Rajabpour, "Thermal expansion coefficient of graphene using molecular dynamics simulation: A comparative study on potential functions"
[7] V. Serdyukov and A. Surtaev, "Experimental study of multi-scale heat transfer characteristics at pool boiling"
[8] A. Tsimpoukis, C. Tanto and D. Valougeorgis, "Time-dependent conductive heat transfer in rarefied polyatomic gases confined between parallel plates"
[9] M. Verdier, D. Lacroix and K. Termentzidis, "Effect of amorphisation around spherical

nanopores on the thermal conductivity of nanoporous silicon"

[10] K. Voitenko, M. Isaiev, A. Pastushenko, D. Andrusenko, A. Kuzmich, V. Lysenko and R. Burbelo, "Thermal transport study across interface nanostructured solid surface/fluid by photoacoustic technique"

[11] X. Zianni, K. Termentzidis and D. Lacroix, "On the dependence of the thermal conductivity of width-modulated nanowires on the number of modulations"

IOP Conf. Series: Journal of Physics: Conf. Series **785** (2017) 011002 doi:10.1088/1742-6596/785/1/011002

Peer review statement

All papers published in this volume of *Journal of Physics: Conference Series* have been peer reviewed through processes administered by the proceedings Editors. Reviews were conducted by expert referees to the professional and scienti c standards expected of a proceedings journal published by IOP Publishing.

IOP Conf. Series: Journal of Physics: Conf. Series **785** (2017) 012001 doi:10.1088/1742-6596/785/1/012001

Thermal transmission at Si/Ge interface: *ab initio* lattice dynamics calculation

A Alkurdi[12] **and S Merabia**[1],

[1] Institut Lumière Matière, UMR5306 Université Claude Bernard Lyon 1-CNRS, Université de Lyon 69622 Villeurbanne Cedex, France

[2] Department of Physics, Al-Baath University, Homs, Syria

E-mail: ali.alkurdi@univ-lyon1.fr, samy.merabia@univ-lyon1.fr

Abstract. We perform lattice dynamics calculations (LD) on silicon/germanium interfaces using *ab initio* interatomic force constants to predict the interfacial phonon transmission as a function of both phonon frequency and the transmission angle. We carry out a spectral and angular analysis to quantify the contribution of each phonon mode in a given scattering direction. The effect of the interaction range was studied at this interface by taking account of more or less atom layers across the interface. Moreover, we were able to predict the thermal boundary conductance (TBC) as a function of the transmission angle and temperature as well. Our results show that, the thermal energy transmission is highly anisotropic while thermal energy reflection is almost isotropic. In addition, we found that it seems there is a global critical angle of transmission beyond which almost no thermal energy is transmitted. This can be used to device high pass phonon filter via changing the orientation of the interface.

1. Introduction

Since the discovery of thermal resistance at the interface between two media by the pioneering experimental work of Kapitza on superfluid helium/solid interfaces [1], many theoretical attempts have been to understand interfacial heat transport. Among the most popular models, the acoustic mismatch model (AMM) [2] and diffuse mismatch model (DMM) [3] have been proposed. The former model AMM consider the phonons as plane waves experiencing specular transmission at the interface so the transmission is a function of the ratio between the impedances of the two media, while in the latter model DMM, all phonons diffusely scatter at interfaces, and the probability of transmission is determined by a mismatch between the densities of states. Both models fail to predict the thermal conductance measured experimentally [4], but they are still widely used because of their simplicity. In the quest for more accurate models to predict thermal boundary conductance (TBC), several methods have been proposed during the last decade, including molecular dynamics [5, 6], Green function [8] and lattice dynamics [7, 10]. Despite its popularity and wide use, molecular dynamics simulations often rely on the consideration of empirical potentials, which do not describe accurately bulk phonon dispersion. Lattice dynamics calculations have been traidtionally devised to give a precise account of bulk phonon dispersion. This level of accuracy has, however, not been reached for describing interfacial heat flow. Let us mention lattice dynamics calculations for one-dimension [9], two-dimension and three-dimension [7] solids. The pioneering work of Young and Maris provided a reasonable agreement with themroreflectance measurements of the Kapitza conductance for interfaces between metals and dielectrics as long as their Debye temperatures differ by less than a factor of 5 [11]. In this article, we perform lattice dynamics calculations using *ab initio* interatomic force constans in order to quantify phonon transmission at silicon/germanium interface. We show that these calculations provide a full description of the interfacial transmission including the frequencies and directions of both transmitted and reflected phonons and enable us to predict the thermal boundary conductance (TBC) at the interface as a function of the transmission angle.

2. Theoretical Model

In this section, we present the main equations used in lattice dynamics calculations (LD) for a perfect interface between two diamond-like structure solids in three-dimensions. Thereby, we can characterize the properties of incident and excited phonons required to calculate the phonon energy transmission at the interface. The interaction in each medium is supposed to be determined by interatomic force constants $\mathbf{\Phi}(\mathbf{r})$ derived from *ab initio* calculations [12]. In the bulk, the equation of motion for an atom j of mass m_j displaced by $\mathbf{u}(\mathbf{r}_j)$ can be read as:

$$m_j\ddot{\mathbf{u}}(\mathbf{r}_j) = -\sum_{\acute{j}} \mathbf{\Phi}(\mathbf{r}_{j\acute{j}})\mathbf{u}(\mathbf{r}_{\acute{j}}) \tag{1}$$

where the sum is over all atoms including atom j itself (these are called self-terms) and $\mathbf{\Phi}$ is the interatomic force constant matrix limited here to the harmonic terms. Eq. 1 admits a general solution of the form :

$$\mathbf{u}(\mathbf{r}_j, t) = \mathbf{A}(\mathbf{k}, \nu) \exp\left[i(\mathbf{k}.\mathbf{r}_j - \omega t)\right] \tag{2}$$

This solution represents a wave of mode p propagating in the $\mathbf{r}$ direction with the amplitude $\mathbf{A} = A\mathbf{e}$. Therefore, Eq. 1 can be rewritten as:

$$\mathbf{D}(\mathbf{k})\mathbf{e}(\mathbf{k}, \nu) = \omega^2(\mathbf{k}, p)\mathbf{e}(\mathbf{k}, \nu) \tag{3}$$

$\mathbf{D}$ is a 6×6 dimension matrix called the dynamical matrix and $\mathbf{e}$ is the polarisation vector indicating the direction of the corresponding atomic displacement vector. The interface is represented in Fig. 1, where we divide the system into three regions: the left lead where the incident phonons come from, the central region corresponding to the location of the interface and the right lead where the transmitted phonons come out, denoted by L, C and R, respectively. The interface is oriented to be perpendicular to the z-axis corresponding to the [001] crystalline direction.

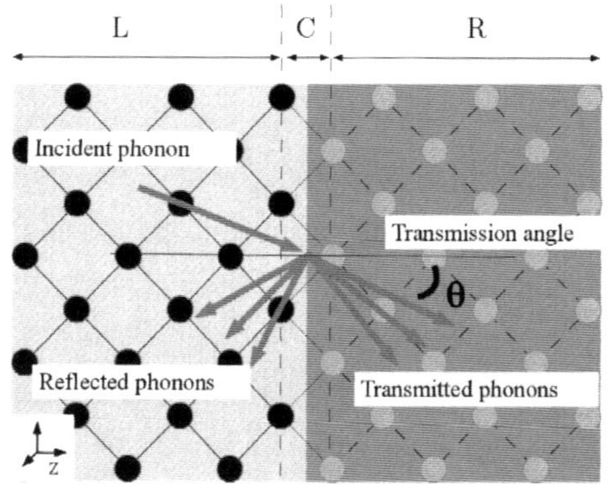

Figure 1. Schematic of the perfect interface between Si/Ge. Coming from the left lead, the incident phonon strikes the interface in the central region and is decomposed into transmitted phonons in the right lead and reflected phonons in the left lead.

After decomposing the dynamical matrix into three or more region-dependent dynamical sub-matrices [10], we are able to determine the excited modes. At the interface, we assume elastic scattering, in other words, the incident phonon can only excite phonons having the same frequency

$$\omega(\mathbf{k}^L) = \omega(\mathbf{k}^R) = \omega(\mathbf{k}) \tag{4}$$

Also, at the interface the components of the wave vector parallel to the interface are conserved [13]

$$k_x^L = k_x^R = k_x \quad \text{and} \quad k_y^L = k_y^R = k_y \tag{5}$$

With these two requirements, we can find the excited modes by solving a generalized linear eigensystem

$$\begin{pmatrix} -\mathbf{D}^{L'} & M\omega^2\mathbf{I} - \mathbf{D}_S^C & -\mathbf{D}^{R'} & -\mathbf{D}^{R''} \\ \mathbf{I} & 0 & 0 & 0 \\ 0 & \mathbf{I} & 0 & 0 \\ 0 & 0 & \mathbf{I} & 0 \end{pmatrix}^{-1} \begin{pmatrix} \mathbf{D}^{L''} & 0 & 0 & 0 \\ 0 & \mathbf{I} & 0 & 0 \\ 0 & 0 & \mathbf{I} & 0 \\ 0 & 0 & 0 & \mathbf{I} \end{pmatrix} \begin{pmatrix} \mathbf{e} \\ \mathbf{e}' \\ \mathbf{e}'' \\ \mathbf{e}''' \end{pmatrix} = \Gamma \begin{pmatrix} \mathbf{e} \\ \mathbf{e}' \\ \mathbf{e}'' \\ \mathbf{e}''' \end{pmatrix} \tag{6}$$

where $\Gamma \equiv \exp(ik_z a/2)$, $\mathbf{e}' = \Gamma\mathbf{e}$, $\mathbf{e}'' = \Gamma^2\mathbf{e}$ and $\mathbf{e}''' = \Gamma^3\mathbf{e}$. The subscript S indicates that the corresponding matrix contains all self-terms. This will give us all possible z-components of the reflected and transmitted wave vectors, but we select only those which have a physical meaning as described by Zhao and Freund [10]. Note also that the order in Γ depends on the number of interacting neighbours that we consider, higher orders correspond to larger interaction range. For some incident wavevectors, the generalized matrix in Eq. 6 may be singular, we solve then this equation by a direct method. We eliminate the cases where $\Gamma = 0$ and $\Gamma = \infty$ which give unphysical values of Γ.
To completely describe the transmission at the interface, we need a set of $6 \times N$ equations, where N is the number of the interacting atomic layers taken into account. The complex displacement vector, $\mathbf{u}(\mathbf{r}_{ij})$ of the ith atom in the jth primitive unit cell in the left lead can be expressed as a superposition of the incident mode and the excited modes in the left lead :

$$\mathbf{u}(\mathbf{r}_{ij}) = \mathbf{A}_i(\mathbf{r}_i, \mathbf{k}, \nu) \exp[i\mathbf{k}.\mathbf{r}_j] + \sum_{\mathbf{k}_r,\nu}^{n_r} \mathbf{A}_r(\mathbf{r}_i, \mathbf{k}_r, \nu) \exp[i\mathbf{k}_r.\mathbf{r}_j] \tag{7}$$

where $\mathbf{A}_r$ is the amplitude of the reflected waves and n_r the number of allowed modes. Similarly, the displacement vector in the outcoming medium (right lead) is

$$\mathbf{u}(\mathbf{r}_{ij}) = \sum_{\mathbf{k}_t,\nu}^{n_t} \mathbf{A}_t(\mathbf{r}_i, \mathbf{k}_t, \nu) \exp[i\mathbf{k}_t.\mathbf{r}_j] \tag{8}$$

The equation of motion of the $l$the atom layer at interface can be written in the general form:

$$\sum_{\mathbf{k}_{i,r,t}} \left[\left(\omega^2\mathbf{I} - \mathbf{D}_s^L - \mathbf{D}_s^{R,\overset{+}{l}} - \mathbf{D}_s^{C,l} - \mathbf{D}_s^{int,\bar{l}} \right) \mathbf{A}_{i,r} - \mathbf{D}_{int}^{R,\bar{l}} \mathbf{A}_t \right] = 0 \tag{9}$$

$$\sum_{\mathbf{k}_{i,r,t}} \left[-\mathbf{D}_{int}^{L,\bar{l}} \mathbf{A}_{i,r} + \left(\omega^2\mathbf{I} - \mathbf{D}_s^R - \mathbf{D}_s^{L,\overset{+}{l}} - \mathbf{D}_s^{C,l} - \mathbf{D}_s^{int,\bar{l}} \right) \mathbf{A}_t \right] = 0 \tag{10}$$

where the sum is over all wave vectors. The first equation Eq. 9 describes the motion of an atom in the incoming medium (left lead). The interactions with the atoms in the left lead are represented by the left dynamical matrix $\mathbf{D}_s^L$ which includes the self-terms. The interactions with the atoms in the right lead across the interface are represented by the right dynamical matrix with interfacial force constant $\mathbf{D}_{int}^{R\bar{l}}$, from which some terms corresponding to up to the lth neighbour layer were subtracted and added to the right matrix dynamic of the other medium in bulk $\mathbf{D}_s^{R,\overset{+}{l}}$. The subscript s is always used to indicate that the matrix includes the self-terms, while int indicates that the force constants are calculated across the interface, and $\bar{l}$ ($\overset{+}{l}$) refers to up to which layer the corresponding terms are subtracted (resp. added). The matrix $\mathbf{D}_s^{int,\bar{l}}$ contains only the self-terms at interface, from this matrix also some terms of up to lth neighbour were subtracted. Similarly, the second equation Eq. 10 describes the motion of an atom in the outcoming medium (right lead) and the notations remain the same. By solving the previous set of equations, we obtain the amplitude of each excited modes $i.e$ reflected A_r and transmitted A_t phonons and then the energy phonon transmission coefficient defined as the fraction of phonon energy transmitted across the interface, can be obtained :

$$\mathcal{T} = \frac{\rho_t}{\rho_i} \sum_{1}^{n_t} \frac{v_{g,t}^z . |A_t|^2}{v_{g,i}^z . |A_i|^2} \tag{11}$$

where $v_{g,t}^z$ are the group velocities of the transmitted phonons in the direction perpendicular to the interface and n_t is the number of transmitted modes.

3. Results and Discussion

We present now the results obtained by performing the LD calculations on Si/Ge interface as described in the previous section. The interatomic interaction was limited to the 8th neighbours, which corresponds to the 4th unit cell (diamond-like crystals have two atoms per unit cell). Fig. 2 shows the transmission and reflection of all the phonon modes in polar coordinates, and for different interaction ranges that goes from the first unit cell (L1) up to the 4th unit cell (L4). The angular axis is taken as the y-axis. Overall, the transmission coefficient displays a maximum value about 1/2 in the normal transmission direction corresponding to $\theta = 0°$ and it decreases when we go to the grazing direction, $\theta = 90°$. From these curves, we can see that the phonon transmission is anisotropic with relative low values lying in the interval of transmission angle between $\theta = 50°$ and $\theta = 90°$. This transmission angle $\theta = 50°$ beyond which the transmission is small can be seen as a global critical angle. Thermal phonon reflection seems to be almost isotropic with maximum value about 0.7 in the direction of 90° of transmission angle. The transmission coefficient is found to be almost insensitive to the interaction range, but the reflection coefficient displays a little increase when the interaction goes up to the 4th unit cell.

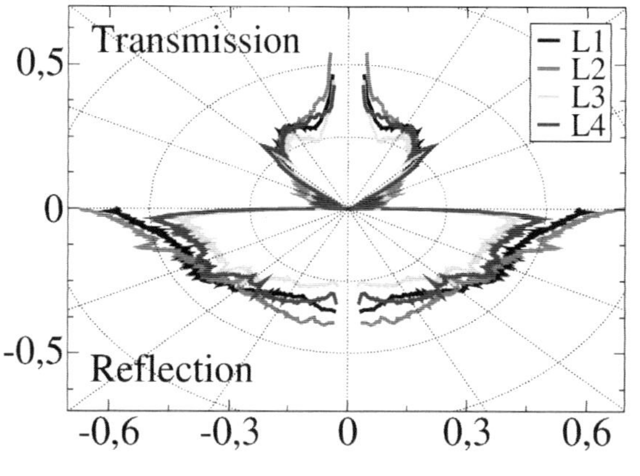

Figure 2. (Color inline) Phonon energy transmission at Si/Ge in polar coordinates for different interaction ranges. L1 refers to the interaction limited to only the nearest unit cell while L4 refers to the interaction up to the 4th unit cell.

We have performed a spectral and angular analysis to quantify the contribution of each phonon mode in a given scattering direction. Fig. 3 depicts this analysis showing that the transmission is highly anisotropic and we found that the probability of transmission is more important in the normal direction up to a transmission angle of $\theta = 50^\circ$. In this orientation, phonons at frequencies higher than 20 rad/ps have a larger probability to be transmitted than phonons having lower frequency. For frequencies lower than 10 (rad/ps), we observe that energy transmission has a limit beyond which the phonon transmission is very low, which confirms the existence of a critical angle at least for these low frequency phonons.

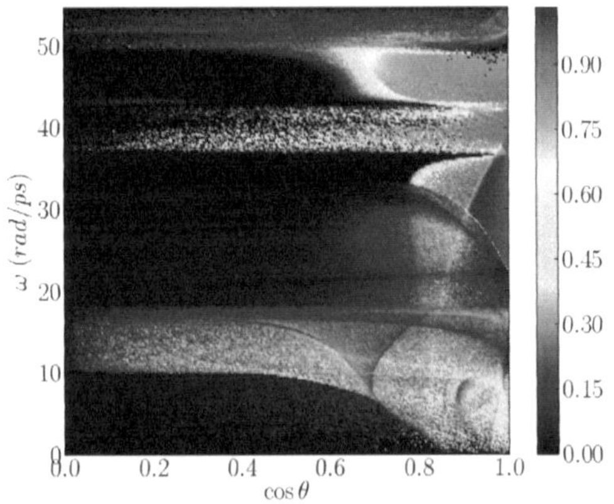

Figure 3. (Color inline) Colormap of the phonon energy transmission at Si/Ge interface as a function of both the angular frequency and the transmission angle. The interaction range was taken up to 2nd unit cell here.

We have also predicted the thermal boundary conductance (TBC) at silicon/germanium interface, which was found to be $G_{\text{Si/Ge}} \simeq 200$ (MW/m^2/K) at room temperature when the interaction is limited to the first nearest unit cell (L1). The same value was predicted by Tian *et al* [14] for ideal Si/Ge interface using Green's function. In Fig. 4, we plotted the cumulative TBC as a function of the transmission angle, and we see that a significant contribution to energy transmission across the interface comes to large extent from angular directions close to normal incidence. This is an expected result as the transmission in these directions has been found to be more important. Note also that the TBC exhibits a significant drop in magnitude when the interaction range goes beyond the first unit cell (L1). This behaviour may be surprising and contrary to the intuition, nevertheless, it proves that enhanced interfacial bonding does not imply better phonon transmission. Therefore, we can conclude that to predict accurately the TBC, it is mandatory to consider at least three nearest unit cells, and that popular short range potentials describing diamond like crystals such as the Stillinger-Weber or the Tersoff potentials, will do a poor job in predicting interfacial heat transfer. Unfortunately, a direct comparison with experiments is not possible since we are not aware of an experimental data reporting the TBC of a single Si/Ge interface.

4. Conclusion

In summary, we have used LD calculations together with *ab initio* interatomic force constans to quantify phonon transmission at silicon/germanium interface and predict the thermal boundary conductance. These calculations have allowed us to quantify the transmission and reflection coefficients as a function of transmission angle, enlighting the important contribution of the phonons which are transmitted in the normal direction. A spectral and angular analysis of phonon transmission has been done to elaborate a complete view on the contribution of each phonon mode in a given scattering direction. Finally, we have emphasized the necessity to take into account at least three interacting unit cells to properly describe the TBC at Si/Ge interfaces. This work may be extended in different ways. First, it may be employed to predict

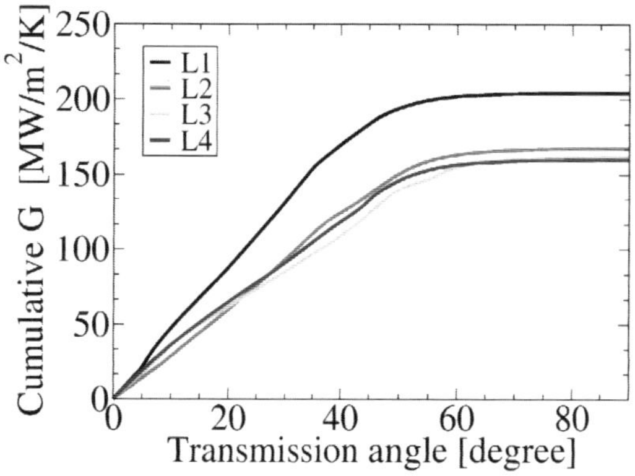

Figure 4. (Color inline) The cumulative TBC at room temperature for Si/Ge interface for different interaction ranges.

phonon scattering at rough interfaces following the lines of [15]. This may shed light on the possibility to enhance heat conduction through periodic patterned interfaces as considered in [16]. Another extension concerns the possibility to build multiscale models of phonon transport across solid multilayers and superlattices. To this end, the phonon transmission coefficient calculated here from first principles calculations may serve as input in a Monte Carlo simulation of the Boltzmann Transport Equation, which generally employs empirical transmission coefficient method to describe phonon scattering at interfaces. This will permit to simulate heat transport in devices having realistic sizes still keeping a microscopic description of interfacial conduction.

5. Acknowledgments

The authors acknowledge financial support from the ANR Mascoth funded by the Agence Nationale de la Recherche.

References

[1] P.L. Kapitza, *J. Phys. (USSR)*, **4** 181 (1941)
[2] W.A. Little, *Can. J. Phys. 37* 334 (1959)
[3] E.T. Swartz and R.O. Pohl, *Rev. Mod. Phys.* **61** 605 (1989)
[4] D.G. Cahill, W.K. Ford, K.E. Goodson, G.D. Mahan, A. Majumdar, H.J. Maris, R. Merlin and S.R. Philpot, *J. App. Phys.* **93** 793 (2003)
[5] E.S. Landry and A.J.H. McGaughey, *Phys. Rev. B* **80** 165304 (2009)
[6] S. Merabia and K. Termentzidis, *Phys. Rev. B* (2012) **86** 094303
[7] D.A. Young, and H.J. Maris, *Phys. Rev. B* **40** 3685 (1989)
[8] W. Zhang, T.S. Fisher and N. Mingo, *AMSE J. Heat Transfer* 129 (2007) 483
[9] L. Zhang, P. Keblinski, J.S. Wang, and B. Li, *Phys. Rev. B* **83** 064303 (2011)
[10] H. Zhao and J.B. Freund, *J. Appl. Phys.* **97** 024903 (2005)
[11] R.J. Stoner and H.J. Maris, *Phys. Rev. B* **48** 16373 (1993)
[12] M. Aouissi, I. Hamdi, N. Meskini and A. Qteish, *Phys. Rev. B* **74** 054302 (2006)
[13] S. Pettersson and G.D. Mahan, *Phys. Rev. B* **42** 7386-7390 (1990)
[14] Z. Tian, K. Esfarjani, and G. Chen, *Phys. Rev. B* **86** 235304 (2012)
[15] Zhao, H. and Freund, J. B., *J. Appl. Phys.* **105** 013515 (2009)
[16] S. Merabia and K. Terzmentzidis, *Phys. Rev. B* **89** (2014) 054309

Inverse heat conduction problem in a phase change memory device

Jean-Luc Battaglia[1], Indrayush De[1], Véronique Sousa[2]
[1]I2M Laboratory, University of Bordeaux, UMR CNRS 5295, 351 cours de la liberation 33405 Talence cedex, France
[2] CEA-LETI, MINATEC, 17 rue des Martyrs, F-38054 Grenoble Cedex 9, France

E-mail: jean-luc.battaglia@u-bordeaux.fr

Abstract. An invers heat conduction problem is solved considering the thermal investigation of a phase change memory device using the scanning thermal microscopy. The heat transfer model rests on system identification for the probe thermal impedance and on a finite element method for the device thermal impedance. Unknown parameters in the model are then identified using a nonlinear least square algorithm that minimizes the quadratic gap between the measured probe temperature and the simulated one.

1 Introduction

In a previous study [1], we demonstrated that the scanning thermal microscopy (SThM) is a very efficient tool to perform the thermal investigation of a phase change memory device at the microscale. In previous studies, the thermal properties of each material involved in a phase change memory as well as the thermal boundary resistances (TBR) at the interfaces between those materials have been measured in specific configurations, generally as thin films [2-4]. It is then questionable about the reliability of those properties, mainly at the level of the TBR, considering the device itself. The goal of this paper is to implement an inverse approach in order to identify the parameters that are involved in the model for the heat transfer within the SThM experiment.

The phase change material (PCM) is the $Ge_2Sb_2Te_5$ (GST) very well known alloy [6] that shows a rapid and reversible phase-change switching from amorphous (α) to face centered cubic (*fcc*) metastable and hexagonal close packed (*hcp*) stable crystalline phases. The amorphous-to-crystalline transition is achieved by heating the GST above the glass transition temperature that is 130°C for α to *fcc* transition while it is around 300°C for the *fcc* to *hcp* transition. A remarkable, and one of the most interesting features concerning the practical implementation of the PCM in microelectronics, is that the GST electrical resistance varies along decades according to the crystalline state [7]. The practical functioning of the device is described in [1]. During the past ten years, PCM attracted large attention thanks to their maturity and their amazing performances. Nevertheless, numerous challenges remain regarding the bit retention time at elevated temperature that implies a highest thermal stability of the amorphous state at high temperature, the fastest transformation speed with a low pulse current, the cyclability and finally the scalability implying that the switching properties have to be maintained at dimensions well below tenth of nanometers. With those objectives, it has been shown that confinement of phase change materials, in a μ-trench for instance as presented in the paper of Lai et *al.* [8], enhances Joule's effect efficiency and highly reduce the reset current. However, we have to mention that other solutions are envisaged as that of implementing the phase change material as a nanowire with very low cross-section area [9]. The device developed by CEA-LETI (*see* Fig. 1) is constituted around a cylindrical trench of crystalline GST surrounded by silicon nitride. The layer under the GST is the classical titanium nitride layer that will play the role of the bottom electrode, namely the heater. The PCM is surrounded by silicon nitride (SiN) that acts as the electrical and thermal insulator, with the additional feature to limit the thermal cross-talk between neighbourhood memory cells. The thickness of the GST cylindrical wall is 30 nm and its thickness at the bottom is 60 nm. Thermal properties of the materials at room temperature have been the subject of previous studies and they have been reported in Tab. 1.

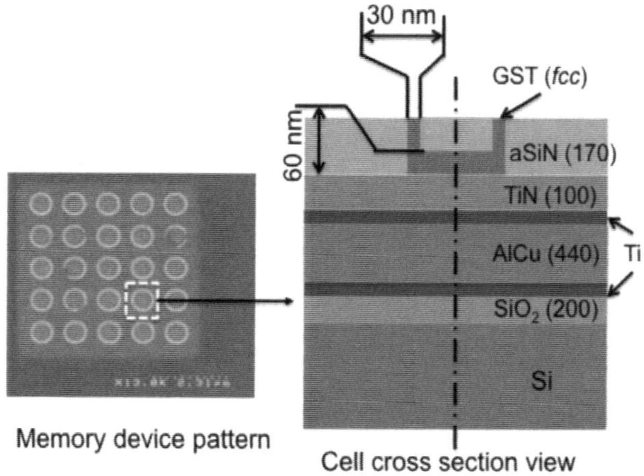

Figure 1. At the left: small area of the memory device pattern; At the right: cross-section of one phase change memory cell (cylindrical revolution). Device developed at CEA-LETI.

Table 1. Thermal properties of materials involved in the phase change memory cell (k: thermal conductivity, ρ: density and C_p: specific heat, Θ_D : Debye temperature, v_T: phonon transverse velocity, v_L: phonon longitudinal velocity) at room temperature.*Cu content being very low in the AlCu alloy, data for Al have been considered.

Material	k (W.m^{-1}.K^{-1})	ρ (kg.m^{-3})	C_p (J.kg^{-1}.K^{-1})	Θ_D (K)	v_T (m.s^{-1})	v_L (m.s^{-1})
Si	131	2329	700	-	-	-
SiO_2 [10]	1.4	2200	787	-	-	-
AlCu*	238	2700	900	-	-	-
Ti [5]	13	4500	540	-	-	-
TiN [5]	33.6	5400	220	580	5110	10200
GST (α) [3]	0.18	5870	218	136	1350	2250
GST (*fcc*) [3]	0.45	6270	205	197	1914	3190
a-Si_3N_4 [10]	2.1	2900	400	985	6200	10300

The thermal boundary resistance (TBR) at the different interfaces between the GST and the surrounding material are more or less well known. They cannot be ignored since a fast calculus shows that they are comparable to the ratio between the thickness and the thermal conductivity of the GST. The TBR at the interface between the GST and the TiN has been the subject of several studies and it was found that it varies significantly according to the GST state. In addition, it was found to be higher that the prediction based on the diffuse mismatch model (DMM) whose asymptotic expression for the high temperature in the Debye approximation is [11]: $R^{\rm DMM}_{\rm TiN\text{-}SiN} = 4/\tau_{\rm SiN,TiN}\,\rho_{\rm TiN}\,C_{p,\rm TiN}\,v_{\rm TiN}$, where $\tau_{\rm SiN,TiN} = \sum_{j=1}^{3} v^{-2}_{g,j,\rm SiN} \Big/ \left(\sum_{j=1}^{3} v^{-2}_{g,j,\rm SiN} + \sum_{j=1}^{3} v^{-2}_{g,j,\rm TiN}\right)$ denotes the phonon transmission coefficient that is calculated from the longitudinal and transverse phonon velocities for the two materials separated by the interface. The TBR at the interface between GST and SiN has not been measured so far to our knowledge. Using the DMM leads to a relatively low TBR as $5\times10^{-9}\,\rm K\cdot m^2\cdot W^{-1}$. However, the Debye temperature for SiN is larger than that of TiN that is already high compare to that of GST. Therefore, a higher TBR at the GST-SiN interface is reasonably expected.

2 Scanning Thermal Microscopy (SThM) technique

The SThM has been implemented within the 3ω mode with the frequency ranging from the DC up to 5 KHz [12]. Anasys Comp. provided the commercial probe that is a Si_3N_4 cantilever on which is deposited a palladium (Pd) strip at the tip. The curvature radius of the apex is about 100 nm. We used a commercial AFM provided by Nanosurf Company. The experimental setup is rather classical although lot of caution have to be put on the electronics that allows extracting and measuring the third harmonic of the voltage drop that is related to the probe temperature thanks to the thermal coefficient of the Pd.

3 Heat transfer model and sensitivity analysis

A model of the heat transfer in the SThM experimental configuration is built and represented on Fig. 2 using the thermal impedances formalism [13]. At each frequency f, the thermal impedance $Z_p(\omega_2)=A_p(\omega_2)\exp(i\delta_p(\omega_2))$, with $\omega_2=4\pi f$, of the probe is deducted from the measurement of the amplitude $A_p(\omega_2)$ and the phase $\delta_p(\omega_2)$ when the probe is out-of-contact from the surface of the sample. Indeed, we consider that the model bias from this "system identification" approach is much smaller than that obtained from an analytical solution based on a simplified geometry of the probe. The electrical resistance of the probe at room temperature is denoted R_0 and the current passing through the Pd wire is i_0. Approaching the probe tip to the surface and ensuring the contact by monitoring the force applied on the cantilever, a very small part of the heat flux generated within the probe can flow the sample. A thermal contact resistance R_c occurs at the interface that has been studied by several colleagues in order to understand all the physical phenomena that are involved regarding the heat transfer at the interface. Among them, the solid-solid contact, the water meniscus formation and the diffusion through air surrounding the contact are well identified. In addition, regarding the method we used to derive the probe thermal impedance, this thermal resistance will also account with the thermal gradient that occurs at the probe tip in the contact mode. In a previous study [5] we observed that the identified thermal contact resistance R_c on calibrated samples did not vary significantly when the thermal conductivity of the material varied from 0.2 to 40 W/m/K. We identified R_c=1.2x10^{-8} m^2.K.W^{-1} from frequency dependent probe temperature measurement (not represented here) when the probe is in contact with the SiN material and we assume that this value remain constant all over the device whatever the material in contact.

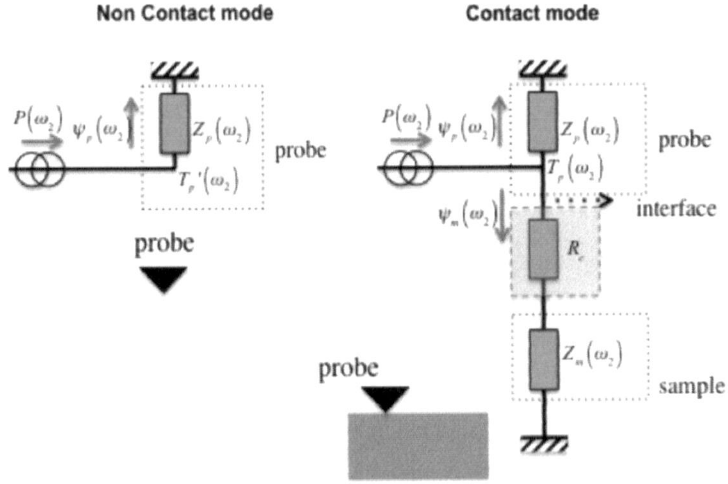

Figure 2. Heat transfer in the Non-Contact and Contact modes using the thermal impedances formalism. The SThM probe used in the present study is a 2-pads contact configuration (photo presented on the lower left). The temperature of the probe is given by $T_p(\omega_2)=Z_T(\omega_2)P(\omega_2)$ with $1/Z_T(\omega_2)=1/(R_c+Z_m(\omega_2))+1/Z_p(\omega_2)$.

Finally, due to the complexity of the domain geometry, the thermal impedance $Z_m(\omega_2)$ for the device has been calculated here using the finite element method. As represented on Fig. 3 the geometry has been first simplified since we verified that, in the working frequency range [100-5000] Hz, the TiN,

AlCu and Ti layers as well as the TBR between those layers can be merged to form the thermal resistance R_i between the SiN and the thermal SiO2 layer at the top of the Si substrate. This former must be considered as a semi-infinite medium within the finite elements technique (node Infinite Domain with COMSOL Multiphysics code). The mesh of the domain appeals to 458096 degrees of freedom (DoF), the system is solved using an iterative multigrid algorithm based on the generalized minimal residual method. The temperature is integrated over the heated area using a 4-order numerical integration technique. As presented also on Fig. 3, a crucial choice has to be made on the value of the characteristic dimension a_{cell} of the cell that is expressed according to the GST tank diameter.

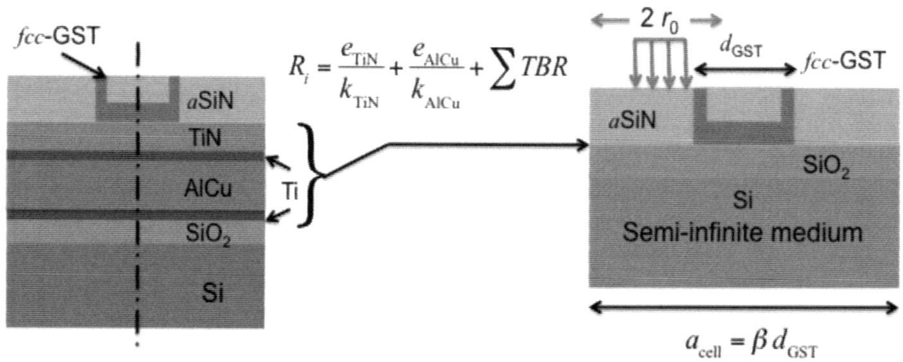

Figure 3. Simplification of the device geometry: the Ti, AlCu and TiN layers are merged to form the thermal resistance R_i, including the TBRs at the interfaces between the layers.

We assumed a contact area between the probe and the surface as a disk; the radius r_0 of the disk was determined using the step technique presented in the paper of Puyoo et *al.* [14]. It is about 100 ± 10 nm and we assumed this value for our configuration. The probe is assumed passing from the SiN onto the GST trench along the x-direction. The x-dependent reduced sensitivity functions of the probe temperature according to the parameter θ are represented in Fig. 4, θ being either the probe thermal impedance Z_p, the geometrical factor β, the TBR at the GST-SiN interface and the thermal conductivity k_{GST} of the PCM have been calculated from the classical finite difference:

$$S^*(x,\theta) = \theta \frac{\mathrm{d}T_p(x,\theta)}{\mathrm{d}\theta} \sim \theta \frac{T_p(x,\theta+\delta\theta) - T_p(x,\theta)}{\delta\theta}$$

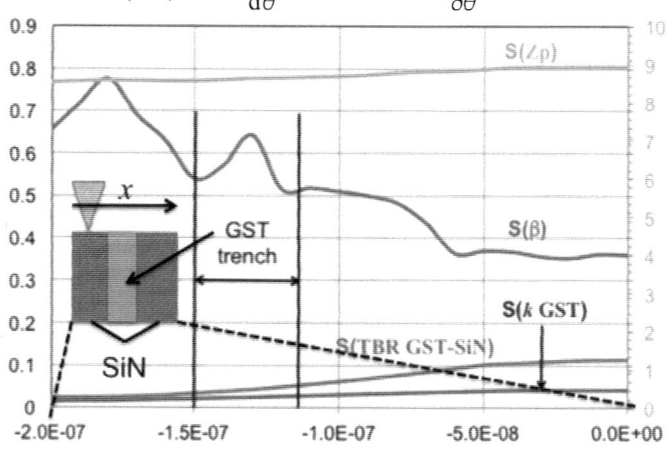

Figure 4. Reduced sensitivity functions of the probe temperature according to the probe thermal impedance Z_p, the geometrical factor β, the TBR at the GST-SiN interface and the thermal conductivity k_{GST} of the PCM.

Obviously we found that the sensitivity on Z_p is huge, meaning that making a small error on this parameter will lead to a very high deviation on the calculated temperature of the probe. Fortunately, we observe that the four sensitivity functions are linearly independent since:

$$\nexists\left(a_0,a_1,a_2,a_3\right)\neq 0 \text{ as: } a_0 S^*\left(Z_p\right)+a_1 S^*\left(\beta\right)+a_2 S^*\left(k_{GST}\right)+a_3 S^*\left(TBR_{SiN-GST}\right)=0$$

Therefore, we proposed to identify the four parameters by minimizing the quadratic gap $J=\sum_x\left(T_p\left(x\right)-\widetilde{T_{p,x}}\right)^2$ between the measured temperature, denoted $\widetilde{T_{p,x}}$, and the calculated one using the Levenberg Marquardt algorithm [15].

4 Results and analysis

Using the experimental data presented already in [1], we performed the identification of the four parameters and we found: $Z_p(523\text{Hz})=(11.10\pm0.025)/P_0\,\text{K.W}^{-1}$; $TBR_{\text{SiN-GST}}=(2.0\pm0.2)\times10^{-8}\,\text{K.m}^2.\text{W}^{-1}$; $k_{GST}=0.46\pm0.1\ \text{W.m}^{-1}.\text{K}^{-1}$; $\beta=3.3\pm0.1$ All the known parameters have been reported in the Tab. 2.

We represented on Fig. 5, the measured and simulated temperature of the probe using the identified parameters. The standard deviations for the parameters have been calculated from the Hessian matrix at the end of the minimization process and the residuals [15]. The value we found for the identified Z_p is exactly the same than that we measured in the out-of-contact mode. As expected we found a value of the TBR at the GST-SiN interface that confirms that the value estimated from the DMM was too low. The GST thermal conductivity is that of the *fcc* state measured when the GST is deposited as a thin film. We observed on Fig. 5 that the change in the temperature probe is only 0.2°C. This comes from the fact that the thermal conductivity of both the amorphous SiN and the GST in *fcc* state are not very different although the amorphous SiN is a little bit more conductive. The increase observed after the GST barrier is obviously due to the heat confinement within the GST micro trench. Considering the thermal properties of the amorphous GST instead of the crystalline ones is clearly visible on the simulated probe temperature on Fig. 5.

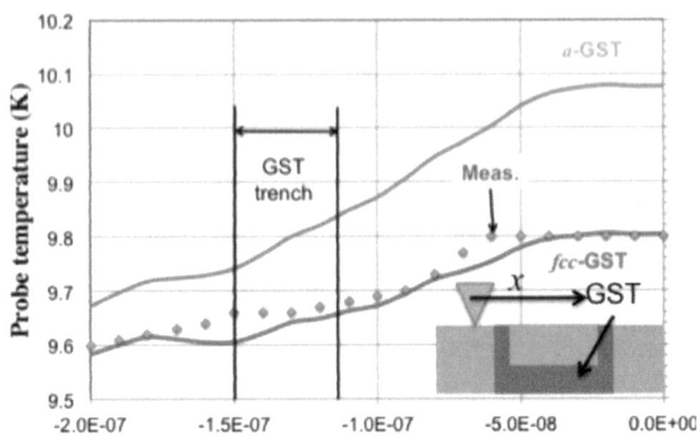

Figure 5. Fit between the experimental data (dots), when the probe follows the direction represented in the inset, and the simulated probe temperature (red line) using the identified values of the four parameters at the end of the minimization process. The simulated probe temperature is calculated from the thermal conductivity of the GST (green line) in order to demonstrate the sensitivity to this parameter for instance.

Table 2. Known parameters.

f	523 Hz	$TBR_{AlCu-TiN}$ (DMM)	$3\text{x}10^{-9}$ m^2.K.W^{-1}
r_0	100 nm	$TBR_{SiO2-AlCu}$ (DMM)	$3\text{x}10^{-9}$ m^2.K.W^{-1}
R_0	125 Ω	$TBR_{Si-SiO2}$ [16]	$5\text{x}10^{-9}$ m^2.K.W^{-1}
R_c	$1.2\text{x}10^{-8}$ m^2.K.W^{-1}	$TBR_{GST-TiN}$ [17-18]	$2\text{x}10^{-8}$ m^2.K.W^{-1}
i_0	800 μA		

5 Conclusion

As it was demonstrated already in previous studies, SThM appears to be reliable to investigate the thermal properties of devices even if the characteristic dimensions of the constitutive elements are lower that the probe spatial resolution. This study shows that the implementation of an inverse approach leads to exploit better the experimental measurements. First, it leads identifying the sensitive thermal parameters in the model that are the thermal conductivity of the PCM (GST), the *TBR* at the PCM-Dielectric (SiN) interface and also additional parameters as the probe thermal impedance and the geometrical domain of investigation. Retrieving the thermal impedance of the probe measured using the out-of-contact mode leads also to be confident with the TBR at the interface between the probe and the surface as well as the contact radius area, that have been both measured in a different configuration. On the other hand, such a characterization allows verifying the reliability of the thermal properties that have been measured in different configurations (as thin films) that than met in the device. However, the main drawback in this approach is related to the CPU time required to complete the identification process. Indeed the minimization algorithm appeals to the simulation of the finite element model a lot of time that makes the process very time consuming. In the present study, only very small amounts of data have been exploited whereas the SThM image is composed from (256x256) pixels. Therefore, further work is designed to reduce strongly the computation times.

Acknowledgements
This work has been funded partly by Conseil Régional d'Aquitaine and the CNRS 102758 project with Epsilon Engineering Company.

References

[1] Battaglia J - L, De I, Saci A, Kusiak A, Sousa V 2016 Journal of Physics: Conference Series **745** 032098
[2] Fallica R, Battaglia J L, Cocco S, Monguzzi C, Teren A, Wiemer C, Varesi E, Cecchini R, Gotti A and Fanciulli M 2009 J. Chem. Eng. Data **54** 1698-1701
[3] Battaglia J - L, Kusiak A, Shick V, Cappella A, Wiemer C, Longo M, Varesi E 2010 J. Appl. Phys. **107** 044314
[4] Battaglia J.-L., Kusiak A, Saci A, Fallica R, Lamperti A, Wiemer C 2015 Appl. Phys. Lett. **105**, 121903
[5] Saci A, Battaglia J L and De I 2015 IEEE Transactions on Nanotechnology **14**(6), 1-4
[6] Lencer D, Salinga M, Grabowski B, Hickel T, Neugebauer J and Wuttig M 2008 Nature Materials 7 972-977
[7] Wong H S P, Raoux S, Kim S, Liang J, Reifenberg J P, Rajendran B, Asheghi M and Goodson K E 2010 Proceedings of the IEEE Vol. **98**, No. 12
[8] S. C. Lai et *al.*, 2013 Symp. VLSI Tech Dig., p.132-133
[9] Lee S – H, Ko D - K, Jung Y, and Agarwal R 2006 Appl. Phys. Lett. **89** 223116
[10] Griffin A J, Brotzen F R, Loos P J 1994 J. Appl. Phys. **76**, 7
[11] Chen G 1998 Phys. Rev. B **57**(23) 14958
[12] Lefèvre S and Volz S 2005 Rev. Sci. Instrum. **76**, 033701
[13] Maillet D, André S, Batsale J-C, Degiovanni A, and Moyne C 2000 *Thermal quadrupoles; solving the heat equation through integral transforms* Wiley edition
[14] Puyoo E., Grauby S., Rampnoux J.-M., Rouvière E., and Dilhaire S., Rev. Sci. Instrum. **81**, 073701 (2010).
[15] Beck J V and Arnold K J 1976 *Parameter Estimation in Engineering and Science* John Wiley & Sons
[16] Lee S M and Cahill D G 1997 J. Appl. Phys. **81**, 6
[17] Reifenberg J P, Kencke D L and Goodson K E 2008 IEEE Electron. Device Lett. **29** 1112
[18] Reifenberg J P, Chang K W, Panzer M A, Kim S, Jeremy A. Rowlette, M. Asheghi, H-S P Wong, and Goodson K E 2010 IEEE Electron Device Letters **31** 1

IOP Conf. Series: Journal of Physics: Conf. Series **785** (2017) 012003 doi:10.1088/1742-6596/785/1/012003

Tayloring energy levels with curvature ? An illustration of Da Costa formalism

Sébastien Fumeron, Bertrand Berche
IJL, UMR Université de Lorraine - CNRS 7198 BP 70239, 54506 Vandœuvre les Nancy, France

Fernando Moraes, Fernando Santos
Universidade Federal de Pernambuco, 50670-901, Recife, PE, Brazil

Abstract. Nanotubes are generally not perfectly straight cylinders and the local curvature can influence transport properties. As shown by da Costa [1], charged particles moving on a curved surface experience an effective potential which modifies their dynamics. In this paper, we solve a one-electron Schrödinger equation in a distorted nanotube with open boundary conditions. We find that the deformations may open bandgaps suggesting their use in the design of nanotube-based electronic devices.

Keywords: Nanotubes, Ballistic transport, Geometric effects

1. Introduction

Single-walled nanotubes (SWNT) are made by rolling up a sheet of graphene into a cylinder. Graphene consists in a 2D honeycomb lattice of carbon atoms and because of its symmetries, one must view it as a hexagonal Bravais lattive with a two-atom basis. The distance between two nearest neighbours carbon atoms is about 0.142 nm. As there is one π-electron per carbon atom that is not involved in a covalent σ bond, there are as many valence electrons than carbon atoms and these electrons can propagate along the nanotube with peculiar properties [2].

This paper is concerned with ballistic electron transport, which is known to be drastically affected by variations of the tube geometry [3, 4] and therefore, of the local curvature. To simplify the study, we will restrict to physical situations where short-range electron-electron interactions have only weak effects: in practice, deviations from the behaviour of non-interacting electrons occur at low temperature and for nanotubes of small transverse size [5].

Our aim is not to make a quantitatively realistic description of deformed nanotubes, but to use a simple independent electron approach in order to access qualitatively the ballistic transport regime, and possibly, understand the kind of impact that curvature may have on electronic band structure. To do so, we will consider deformed nanotubes in the following situations: shrunk nanotube, nanotube with a bump and a wavy peapod. The Schrödinger equation will be solved numerically for a single particle moving in the deformed cylinder, taking into account the modification of the kinetic energy caused by the non-Euclidean geometry and the effective potential induced by curvature. Then we will determine and discuss the transmittances as a function of the injection energy for all cases.

2. Differential geometry of SWNT

In their natural state, nanotubes are not perfectly straight cylinders and a practical way to model 2D carbon lattices is to work in the vanishing lattice spacing approximation: all characteristic lengths (such as de Broglie wavelength of electrons) are considered as large with respect to primitive lattice vectors (typically $a \approx 2.46$ Å). Hence, the discrete lattice appears as a smooth curved surface that is locally homeomorphic to $\mathbb{R}^3$.

For a curved surface parametrized according to $\mathbf{r}\left(x^1, x^2\right) = \left(x\left(x^1, x^2\right), y\left(x^1, x^2\right), z\left(x^1, x^2\right)\right)$, the induced metric tensor on the surface (or first fundamental form) is defined as

$$g_{ij} = \partial_i \mathbf{r}.\partial_j \mathbf{r} = \begin{pmatrix} \mathbf{e}_1.\mathbf{e}_1 & \mathbf{e}_1.\mathbf{e}_2 \\ \mathbf{e}_2.\mathbf{e}_1 & \mathbf{e}_2.\mathbf{e}_2 \end{pmatrix} \tag{1}$$

In this expression, Einstein's summation convention on repeated indices was used, as will also be done in the remainder of this paper. The second quantity of interest is the shape tensor (or second fundamental form) defined as

$$h_{ij} = \begin{pmatrix} \mathbf{N}.\partial_{11}\mathbf{r} & \mathbf{N}.\partial_{12}\mathbf{r} \\ \mathbf{N}.\partial_{21}\mathbf{r} & \mathbf{N}.\partial_{22}\mathbf{r} \end{pmatrix} \tag{2}$$

These two objects are needed for computing the mean and Gaussian curvature scalars M and K, as prescribed by:

$$M = \frac{1}{2}\left(\frac{1}{R_1} + \frac{1}{R_2}\right) = \frac{\det h_{ij}}{\det g_{ij}}, \qquad K = \frac{1}{R_1 R_2} = \frac{1}{2}\mathrm{trace}\left(h_{ij} g^{ij}\right) \tag{3}$$

where R_1, R_2 are the local principal curvature radii and g^{ij} denotes the inverse metric of g_{ij}.

For a graphene sheet presenting one smooth ripple a relevant set of parametric equations is simply found to be $x = \rho(q)\cos\phi, \quad y = q, \quad z = \rho(q)\sin\phi$, where $q \in \mathbb{R}^1$ and $\phi \in \mathbb{S}^1$ (here the axis of the nanotube is chosen to be the y-axis). With such parametrization, the first and second fundamental forms are respectively:

$$g_{ij} = \begin{pmatrix} \rho(q)^2 & 0 \\ 0 & 1+\rho'(q)^2 \end{pmatrix}, \qquad h_{ij} = \frac{1}{\sqrt{1+\rho'(q)^2}} \begin{pmatrix} \rho(q) & 0 \\ 0 & -\rho''(q) \end{pmatrix} \tag{4}$$

3. Quantizing constrained particles on a curved surface

The problem of a free quantum particle moving in a curved surface was solved by R. C. T. da Costa [1] in a seminal paper published in 1981. da Costa approach has ever since been successfully applied to a wide range of two-dimensional systems ([6],[7],[8]) and experimental verification was done in [9] by measuring the high-resolution ultraviolet photoemission spectra of a C_{60} peanut-shaped polymer.

Introducing a confining potential given by an infinite square well in the normal direction, da Costa found that in-plane and out-of-the-plane dynamics decouple (adiabatic approximation) in such a way that the Schrödinger operator describing the particle reduces to [1]

$$H = -\frac{\hbar^2}{2m}\sum_{i,j}\frac{1}{\sqrt{g}}\frac{\partial}{\partial x^i}\left(\sqrt{|g|}g^{ij}\frac{\partial}{\partial x^j}\right) + V_{\mathrm{geo}}, \tag{5}$$

where $g = \det g_{ij}$ and the geometric potential is given by

$$V_{\mathrm{geo}} = -\frac{\hbar^2}{2m}\left(M^2 - K\right). \tag{6}$$

For the parametrization of the SWNT seen before, this potential can be found analytically as:

$$V_{\text{geo}} = -\frac{\hbar^2}{8m}\frac{[1+\rho'(q)^2+\rho(q)\rho''(q)]^2}{\rho(q)^2[1+\rho'(q)^2]^3}. \tag{7}$$

Therefore, to obtain the stationnary states of the Hamiltonian, one must solve:

$$\Psi'' + \frac{\rho'}{\rho}\left[1-\frac{\rho\rho''}{1+(\rho')^2}\right].\Psi' + \left(1+(\rho')^2\right)\left[\frac{2m}{\hbar^2}(E-V_{\text{geo}})]-\frac{\ell^2}{\rho^2}\right]\Psi = 0. \tag{8}$$

E is the total energy, Ψ are the longitudinal eigenfunctions and $l\hbar$ the angular momentum. Taking Robin boundary conditions, solutions are used to compute the probability currents density $j = \frac{\hbar}{2mi}\left(\Psi^*\Psi' - \Psi\Psi^{*'}\right)$ and the transmittance $T = \frac{j_{trans}}{j_{inc}}$. From these expressions, we implemented a MAPLE code, with electron effective mass $m_e = 5.68 \times 10^{-27}\text{meV.s}^2/\text{nm}^2$.

4. Results

As shown in Figs. 1a, 2a and 3a, we considered a SWNT with a single bump, a single pinch and a wavy structure. Taking R as the radius of the straight nanotube, the corrugation was introduced by using

$$\rho(y) = R \pm \frac{\epsilon}{2}\left[1-\cos\left(2\frac{n\pi\, y}{L}\right)\right] \tag{9}$$

to generate the surface of revolution. ϵ is the strength of the deformation, the $+$ sign was used for the single bump and wavy structures, the $-$ sign for the pinched tube. We used $n = 1$ for the pinched/bumped nanotubes, and $n = 2\,,3\,,4\,,5\,,6$ for the wavy structures.

da Costa potential for a pinch and a bump is depicted in Figs. 1b and 2b, respectively. As the bump has only positive Gaussian curvature, the pinched case, due to its deeper potential well, lowers the quasibound energy levels thus lowering the resonance peak positions as compared to the bumped case. Fig. 3 shows the geometric potential and transmittance for wavy structures with varying number of bumps. As the number of bumps increases, the geometric potential starts to look like the Dirac comb. As in the Kroning-Penney problem, periodicity of the potential opens a gap in the energy spectrum which enlarges with the increasing number of oscillations. The effect on the transmittance is seen in Fig. 3. As expected, the energy gap sensibly reduces the transmittance and the effect becomes sharper with the number of bumps.

5. Conclusions

da Costa potential has to be used anytime a particle is constrained onto a surface (graphene sheets and alike) or a curve (rings). It shows that mean and Gaussian curvatures strongly impact ballistic transport of electrons and in particular, transmittance profiles show that periodic corrugation could lead to bandgaps opening [10]. Conversely, it could be a convenient tool to tune outer geometric properties of SWNT in order to obtain a prescribed band structure.

However, for that purpose, several refinements should be added to the present model. Among them, the influence of electron spin and its couplings to the chirality of nantubes is of prime importance. Torsion is not accounted for in da Costa approach and a proper way to implement it is to derive the effective potential starting from Dirac equation on curved spacetime.

References

[1] da Costa R C T 1981 *Phys. Rev. A* **23**(4) 1982–1987
[2] Meyer J C, Geim A K, Katsnelson M I, Novoselov K S, Booth T J and Roth S 2007 *Nature* **446** 60–63
[3] Marchi A, Reggiani S, Rudan M and Bertoni A 2005 *Phys. Rev. B* **72**(3) 035403
[4] Taira H and Shima H 2007 *Surface Science* **601** 5270 – 5275 ISSN 0039-6028 proceedings of the 10th ISSP International Symposium on Nanoscience at Surfaces

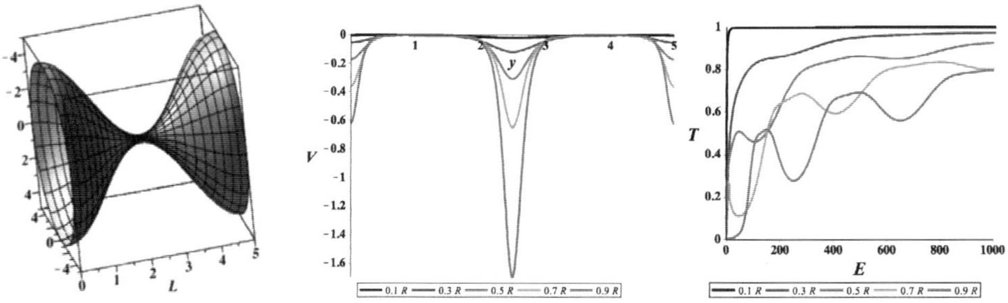

Figure 1: (a) Pinched nanotube. (b) Geometric potential due to the deformation shown in (a) and (c) transmittance as function of incident energy (in meV) for different waist sizes (ϵR).

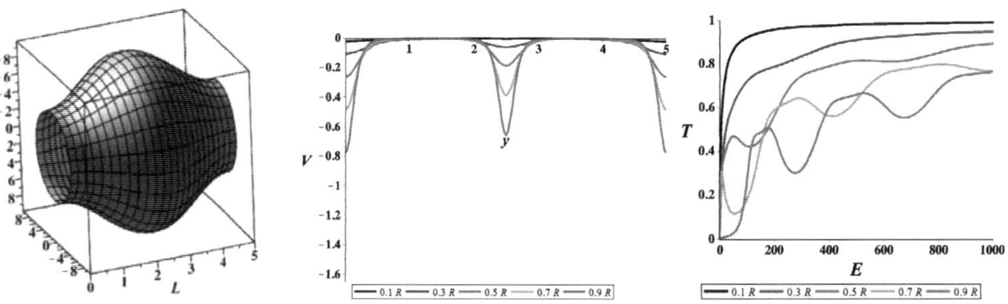

Figure 2: (a) Nanotube with a single bump. (b) Geometric potential due to the deformation shown in (a) and (c) transmittance as function of incident energy (in meV) for different bump sizes (ϵR).

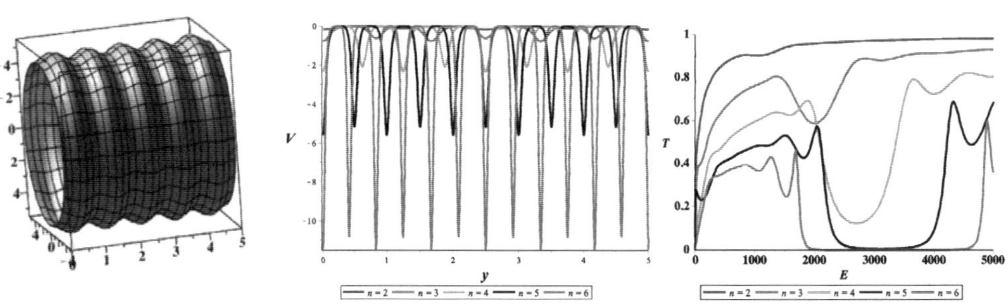

Figure 3: (a) Wavy nanotube. (b) Geometric potential due to the deformation shown in (a) for different numbers of bumps. (c) Transmittance as function of incident energy (in meV).

[5] Kane C, Balents L and Fisher M P A 1997 *Phys. Rev. Lett.* **79**(25) 5086–5089
[6] Chang C H, van den Brink J and Ortix C 2014 *Phys. Rev. Lett.* **113**(22) 227205
[7] Gaididei Y, Kravchuk V P and Sheka D D 2014 *Phys. Rev. Lett.* **112**(25) 257203
[8] Chen K C and Chang C R 2013 *SPIN* **03** 1340006
[9] Onoe J, Ito T, Shima H, Yoshioka H and ichi Kimura S 2012 *EPL (Europhysics Letters)* **98** 27001
[10] Santos F, Fumeron S, Berche B and Moraes F 2016 *Nanotechnology* **27** 135302

IOP Conf. Series: Journal of Physics: Conf. Series **785** (2017) 012004 doi:10.1088/1742-6596/785/1/012004

Modeling of thermal conductivity in high performing thermoelectric materials

E. Hatzikraniotis[1,*], Th. Kyratsi[2], K.M. Paraskevopoulos[1]

(1) Department of Physics, Aristotle University of Thessaloniki, 54124 Thessaloniki, Greece

(2) Department of Mechanical and Manufacturing Engineering, University of Cyprus, 1678 Nicosia, Cyprus

*corresponding author: evris@physics.auth.gr

Abstract. The enhanced TE-performance in Mg_2Si-Mg_2Sn based pseudo-binaries is presented, which is attributed to low thermal conductivity. Sn-Si alloying, reduces the lattice thermal conductivity due to mass fluctuation. Furthermore, miscibility gap in the Sn-Si substitution causes the formation of composites, with Si-rich and Sn-rich phases, which span from mm to nm scale, and these nano-inclusions reduce further lattice thermal conductivity.

1. Introduction

The world's demand for energy is increasing, along with environmental impact. In fact, the majority of the energy actually produced by man is lost, as waste heat. Only an average of about 1/3 of the energy produced by thermal and nuclear power plants, and similarly, only 30% of the energy produced in the internal combustion engine is converted to useful work [1]. If only a portion of the enormous amount of unused waste heat, generated by domestic heating, automotive exhaust, and industrial processes could be recovered, it would cause a tremendous benefit in energy demands. Thermoelectricity (TE) could be of great use in energy harvesting [2, 3].

The thermoelectric performance is ranked by the figure of merit ZT ($\boldsymbol{ZT=S^2\sigma T/\kappa}$), where $\boldsymbol{S}$ is the Seebeck coefficient, $\boldsymbol{\sigma}$ the electrical conductivity, $\boldsymbol{\kappa}$ the total thermal conductivity and $\boldsymbol{T}$ the absolute temperature. A highly-performing TE material should have high $\boldsymbol{S}$ and $\boldsymbol{\sigma}$ and low $\boldsymbol{\kappa}$.

Amongst the various promising materials used in the field of thermoelectrics, Mg_2Si-based compounds have attracted much attention during the last years due to their high ZT in the mid-temperature range, combined to the fact that they are environmentally friendly and they contain widely abundant constituent elements, such as Mg, Sn and Si, with low cost and low density, thus meeting the requirements for large scale production [4]. Tani & Kido [5] presented a ZT as high as 0.86 for Bi-doped Mg_2Si, which seems to be the best ZT reported for the binary system so far. When alloying with Sn, the ternary system $Mg_2Si_{1-X}Sn_X$ presents even higher performance (ZT up to 1.4), due to its lower thermal conductivity that mainly originates from the Si/Sn mass fluctuation in the lattice [6, 7].

In this work we explore thermal conductivity in Sb-doped $Mg_2Si_{0.6}Sn_{0.4}$, by modeling the various mechanisms that act cumulatively increasing the phonon scattering.

2. Experimental

Sb-doped $Mg_2Si_{0.6}Sn_{0.4}$–type materials were synthesized by high purity starting elements, as described in ref. 8. Materials were characterized by powder XRD (PXRD) and found as of pure phase, without any traces of un-reacted Mg, Si, Sn. All major peaks indicate the nominal composition; however they are asymmetric (Fig.1a), with a shoulder at higher 2θ. In fact, $Mg_2Si_{0.6}Sn_{0.4}$ is a composite material which consists of Si-rich & Sn-rich phases in a matrix phase, as has been observed in other members of the same family [7, 9], formed due to the miscibility gap, and span from mm-range down to nm-range [9], as shown in Fig.1b. PXRD diagrams were taken by a Rigaku SmartLab System, while TEM observations were carried out using a JEOL 120CX microscope operating at 100 kV. Thermal conductivity was measured using a laser flash technique on a NETZSCH LFA-457 apparatus, as described in ref. 8. The electrical conductivity and the Seebeck coefficient were determined simultaneously using an ULVAC ZEM-3 (Sinku-Riko) instrument.

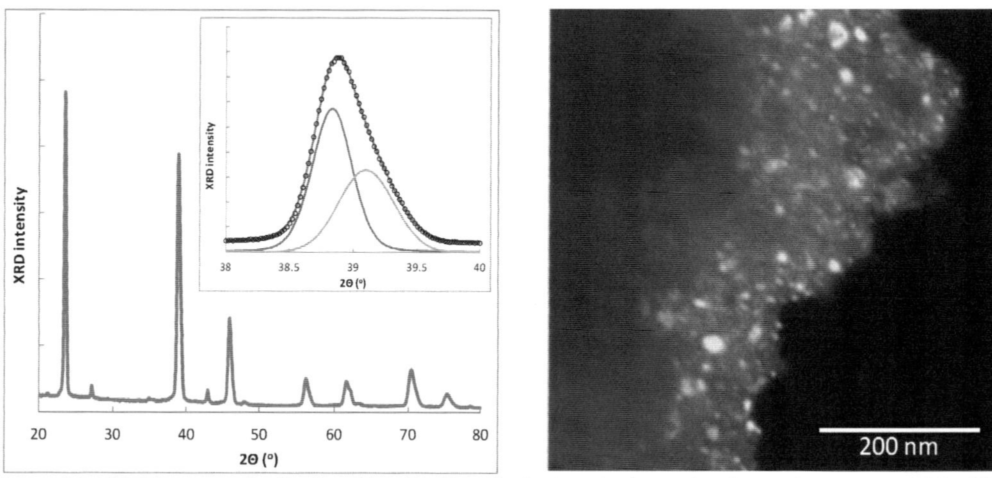

Figure 1 *(a) PXRD pattern for Sb-doped $Mg_2Si_{0.6}Sn_{0.4}$. In the inset is shown the observed asymmetry of the (220) peak. (b) Dark field TEM micrograph with several nano-scale inclusions (white spots).*

3. Results and discussion

Thermal conductivity consists of two additive parts ($\boldsymbol{\kappa=\kappa_L+\kappa_E}$), the lattice ($\boldsymbol{\kappa_L}$) and the electronic ($\boldsymbol{\kappa_E}$) contribution. The electronic contribution to thermal conductivity is given by $\boldsymbol{\kappa_E=L\sigma T}$, where $\boldsymbol{L}$ is the Lorenz number and $\boldsymbol{\sigma}$ the electrical conductivity. The Lorentz number ($\boldsymbol{L}$) was determined from Seebeck coefficient measurements ($\boldsymbol{S}$), through the reduced Fermi level ($\boldsymbol{\eta=E_F/k_BT}$) assuming acoustic phonon scattering:

$$L=\left(\frac{k_B}{e}\right)^2\left[\frac{3F_0(\eta)F_2(\eta)-4F_1^2(\eta)}{F_0^2(\eta)}\right] \quad , \quad S=\pm\frac{k_B}{e}\left[\frac{2F_1(\eta)}{F_0(\eta)}-\eta\right] \tag{1}$$

where $\boldsymbol{F_j(\eta)}$ is the Fermi integral. The resulting lattice thermal conductivity is presented as experimental points in Figure 2a.

3.1. Modeling lattice thermal conductivity

To understand the mechanisms of phonon scattering, semi-classical theoretical calculation based on the modified Callaway's model [10] is conducted:

$$\kappa_L(T)=\frac{k_B}{2\pi^2 u}\left(\frac{k_B T}{\hbar}\right)^3\int_0^{\theta_D/T}\tau_C(x)\frac{x^4e^x}{\left(e^x-1\right)^2}dx \tag{2}$$

IOP Conf. Series: Journal of Physics: Conf. Series 785 (2017) 012004 doi:10.1088/1742-6596/785/1/01

where $\boldsymbol{k_B}$ is the Boltzmann's constant, $\hbar$ is the Plank constant, $\boldsymbol{T}$ is the absolute temperature, $\boldsymbol{u}$ is the average phonon-group velocity, $\boldsymbol{\theta_D}$ the Debye temperature, $\boldsymbol{x=h\omega/k_BT}$, and $\boldsymbol{\tau_C}$ is the phonon relaxation time, which, in our model, combines the scattering from Umklapp/Normal processes ($\boldsymbol{\tau_{UN}}$), nano-structured precipitates ($\boldsymbol{\tau_{NP}}$), alloying ($\boldsymbol{\tau_A}$) and grain boundaries ($\boldsymbol{\tau_B}$) by the equation [11, 12]

$$\frac{1}{\tau_C(T,\omega)} = \frac{1}{\tau_B} + \frac{1}{\tau_{NP}(\omega)} + \frac{1}{\tau_A(\omega)} + \frac{1}{\tau_{UN}(T,\omega)} \tag{3}$$

Some of the mechanisms depend only on geometrical features ($\boldsymbol{\tau_B}$), other on phonon frequency ($\boldsymbol{\omega}$) while Umklapp on both temperature and phonon frequency.

Phonon scattering due to interfaces and _grain boundaries_ ($\boldsymbol{\tau_B}$) is given by [13]:

$$\tau_B^{-1} = \frac{4}{3}\frac{u}{L}\frac{1-t}{t} \quad or \quad \tau_B^{-1} = \frac{u}{L_{eff}} \quad , \quad L_{eff} = \frac{3}{4}\frac{t}{1-t} \tag{4}$$

where $\boldsymbol{t}$ is the phonon transmissivity and $\boldsymbol{L}$ is the average grain size, or scaled to $\boldsymbol{L_{eff}}$.

When _alloying_, the solute atom scattering is accounted by the effective medium approach using a Rayleigh-like expression as described by Klemens [14]:

$$\tau_A^{-1}(\omega) = A\omega^4 \quad , \quad A = \frac{\delta^3}{4\pi u^3}\left[x(1-x)\left\{\left(\frac{\Delta M}{M}\right)^2 + \varepsilon\left(\frac{\Delta\delta}{\delta}\right)^2\right\}\right] \tag{5}$$

where $\boldsymbol{\delta}$ is the radius of the solute atom, $\boldsymbol{M}$ is the molar host weight, $\boldsymbol{\Delta M/M}$ is the mass fluctuation, $\boldsymbol{\Delta\delta/\delta}$ is the strain field fluctuation and ε an adjustable parameter.

The _Umklapp scattering_, derived from 2nd order perturbation is usually given by [15]:

$$\tau_U^{-1}(\omega,T) = B\omega^2 T \exp\left(\frac{-T_D}{3T}\right) \quad , \quad B = \frac{\hbar\gamma^2}{Mu^2 T_D} \tag{6}$$

where γ is the Gruneisen parameter and $\boldsymbol{M}$ the average mass. For normal phonon–phonon scattering, several models were proposed [16], however, it is generally accepted that $(\tau_N)^{-1} \sim \beta(\tau_U)^{-1}$, where $\boldsymbol{\beta}$ is the ratio of two process scattering and is assumed to be temperature independent [12]. Therefore we treat both mechanisms by the same expression, as $\tau_{UN}=\tau_U/(\beta+1)$. The value of parameter $\boldsymbol{\beta}$ is determined by fitting experimental data of pure $Mg_2Si_{0.6}Sn_{0.4}$.

3.2. The effect of nano in thermal conductivity

Relaxation time of phonon scattering due to the nano-scale inclusions is given by a Mathiessen-type combination of short-wavelength and long-wavelength scattering [17]:

$$\tau_N^{-1}(\omega) = u\left(\sigma_S^{-1} + \sigma_L^{-1}\right)^{-1} v_N$$
$$\sigma_S = 2\pi R^2 \quad , \quad \sigma_L(\omega) = \frac{4\pi R^2}{9}\left(\frac{\Delta\rho}{\rho}\right)^2\left(\frac{\omega R}{u}\right)^4 \tag{7}$$

where $\boldsymbol{v_N}$ is the density of the nano particles, $\boldsymbol{\sigma_S}$ & $\boldsymbol{\sigma_S}$ are the short wavelength (geometrical) and long wavelength (Rayleigh-type) scattering cross section (geometrical), $\boldsymbol{R}$ is the average particle radius, $\boldsymbol{\rho}$ is the medium density and $\boldsymbol{\Delta\rho}$ is the density difference between the particle and matrix materials.

Figure 2 shows the effect of nano-structuring in thermal conductivity. At low temperatures, thermal conductivity is increasing, due to the increasing nature of the integral in eq. 2, reaches a maximum, and then decays to a 1/T law due to Umklapp scattering. Standard material properties, like density, Debye temperature, speed of sound, Gruneisen parameter, etc, were taken by linear interpolation between the corresponding values of the two end member materials (Mg_2Si & Mg_2Sn), received from literature [18]. Material characteristics like the average grain size (L=2μm) and average size of nano-particles (2R=22nm) were taken from SEM and TEM measurements [8, 9]. The nano-particle density was roughly estimated by the dark field TEM image and was further refined by adjusting it as the theoretical curve fits the experimental data.

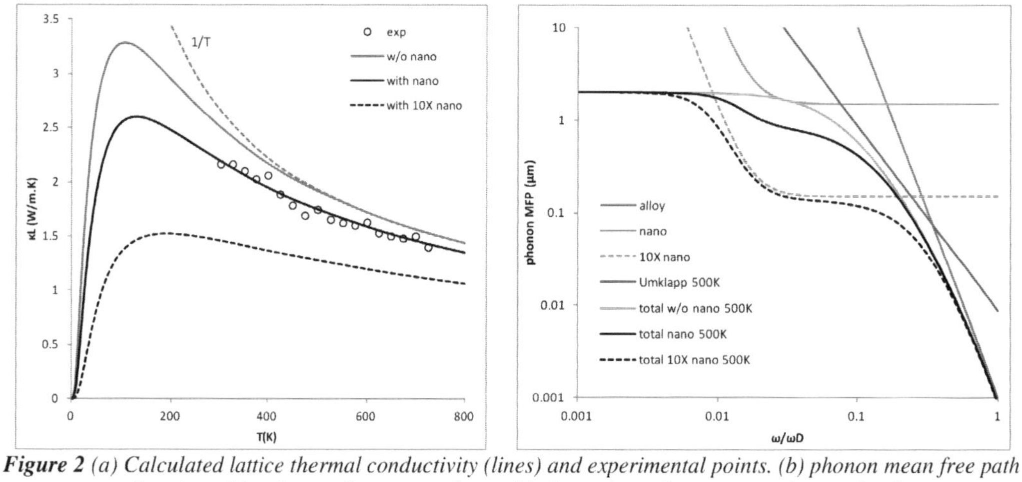

Figure 2 *(a) Calculated lattice thermal conductivity (lines) and experimental points. (b) phonon mean free path as a function of the phonon frequency, along with the various phonon scattering mechanisms.*

Fig.2a shows the effect of nano: Based on the above described model, the top solid curve is without encountering nano, the middle one with nano and the lower (dashed) with 10X density of nano particles. Depending of the nano density, the characteristic peak in lattice thermal conductivity is less and less profound; while the relaxing part (after the peak) considerably deviates from the 1/T law. Clearly, the experimental points are below the top curve (without nano) indicating that an additional phonon scattering mechanism is required, and is better described by the middle curve (with nano).

Fig.2b shows the phonon mean free path (MFP) at 500K, as a function of phonon frequency (**ω/ω_D**). The various acting mechanisms that are reducing the lattice thermal conductivity are shown in figure caption. At low frequencies, phonon MFP is limited by the grain boundary (GB) scattering, while at high frequencies by a combination alloying & Umklapp processes. In the absence of nano, the total phonon MFP is smoothly changing from the GB scattering to alloy/Umklapp scattering as shown by the orange curve. However, when nano is encountered (green curve), a considerable amount of mid-range phonons are scattered by nano inclusions (solid black curve). When nano density is increased by X10, the "nano-curve" (dotted green line) considerably filters out a much larger portion of mid-range phonons, resulting in a significantly lower lattice thermal conductivity.

4. Summary

In this work we model thermal conductivity in Sb-doped $Mg_2Si_{0.6}Sn_{0.4}$, a high performing thermoelectric material. Several mechanisms have been encountered that act cumulatively limiting the phonon mean free path. Standard material properties, were taken by linear interpolation between the corresponding values of the two end member materials (Mg_2Si & Mg_2Sn), received from literature. Analysis has shown that lattice contribution to thermal conductivity can be effectively modeled only if nano-inclusions are taken into account. These nano-inclusions are Si-rich and Sn-rich phases which span from mm to nm scale and reduce further lattice thermal conductivity.

5. Acknowledgement
This work was partially supported by ThermoMag Project, which was co-funded by the European Commission in the FP-7 (contract NMP4-SL-2011-263207) by the European Space Agency (ESA) and by the individual partner organizations.

References

[1] Energy and environment report No 6/2008, European Environment Agency, (www.eea.europa.eu).
[2] S.Scherrer, H.Scherrer, CRC Handbook of Thermoelectrics, CRC Press, Boca Raton (1995) chapter 19-1.
[3] D.M. Rowe (2012), AIP Conference Proceedings 1449 (Proceedings on the 9th European Conference on Thermoelectrics) 485.
[4] J. Tani, M. Takahashi, H. Kido, IOP Conf. Series: Materials Science and Engineering 18 (2011) 142013.
[5] J. Tani, H. Kido, Phys B, 364 (2005) 218.
[6] W. Liu, Q. Zhang, K. Yin, H. Chi, X. Zhou, X. Tang, C. Uher, J Solid State Chem 203, (2013) 333.
[7] (a) A.U. Khan, N.V. Vlachos, Th. Kyratsi, Scripta Materialia 69 (2013) 606–609; (b) A.U. Khan, N.V. Vlachos, E. Hatzikraniotis, G.S. Polymeris, Ch.B. Lioutas, E.C. Stefanaki, K.M. Paraskevopoulos, I. Giapintzakis, Th. Kyratsi, Acta Materialia 77 (2014) 43.
[8] N. Vlachos, G.S. Polymeris, M. Manoli, A.U. Khan, E. Hatzikraniotis, Ch.B. Lioutas, E.C. Stefanaki, E. Pavlidou, K.M. Paraskevopoulos, I. Giapintzakis, and Th. Kyratsi, submitted for publication (2016).
[9] G.S. Polymeris, N. Vlachos, A.U. Khan, E. Hatzikraniotis, Ch.B. Lioutas, A. Delimitis, E. Pavlidou, K.M. Paraskevopoulos, Th. Kyratsi, Acta Materialia 83 (2015) 285.
[10] J. Callaway, Phys. Rev. 113 (1959) 1046.
[11] S-H. Lo, J. He, K. Biswas, M.G. Kanatzidis, V.P. Dravid, Adv. Funct. Mater. (2012), DOI: 10.1002/adfm.201201221
[12] J.He, S.N. Girard, M.G. Kanatzidis, V.P. Dravid, Adv. Funct. Mater. 20 (2010) 764.
[13] G. Chen, T. Zeng, T. D. Borca-Tascius Song, Mater. Sci. Eng. 292 (2000)155.
[14] P.G. Klemens, Phys. Rev. 119 (1960) 507.
[15] D. T. Morelli, J. P. Heremans, Phys. Rev. B 66 (2002) 195304.
[16] L. Yang, J. Wu, L. Zhang, Chin. Phys. 13 (2004) 0516.
[17] W. Kim, S. L. Singer, A. Majumdar, J. M. O. Zide, D. Klenov, A. C. Gossard, S. Stermer, Nano Lett. 8 (2008) 2097.
[18] O. Madelung, U. Rössler, M. Schulz (ed.), Landolt-Börnstein - Group III Condensed Matter 41C: Non-Tetrahedrally Bonded Elements and Binary Compounds I, (1998)

IOP Conf. Series: Journal of Physics: Conf. Series **785** (2017) 012005 doi:10.1088/1742-6596/785/1/012005

Recent improvements on micro-thermocouple based SThM

TP Nguyen[1], L Thiery[1], D Teyssieux[1], D Briand[2] and P Vairac[1].
[1] FEMTO-ST (UMR 6174), Université de Bourgogne Franche-Comté, CNRS, ENSMM, 15B avenue des Montboucons, F-25030 Besançon, France
[2] The Sensors, Actuators and Microsystems Laboratory, EPFL, Neuchâtel, Switzerland

Corresponding Author: tranphong.nguyen@femto-st.fr

Abstract. The scanning thermal microscope (SThM) has become a versatile tool for local surface temperature mapping or measuring thermal properties of solid materials. In this article, we present recent improvements in a SThM system, based on a micro-wire thermocouple probe associated with a quartz tuning fork for contact strength detection. Some results obtained on an electrothermal micro-hotplate device, operated in active and passive modes, allow demonstrating its performance as a coupled force detection and thermal measurement system.

1. Introduction

Probe improvements and the full comprehension of heat transfer at micro/nano-scales are still in progress and remain a major objective for different groups and laboratories for which quantitative thermal microscopy systems remain challenging. SThM measurements can be divided in two groups. The first concerns the measurement of the temperature fields of a surface and the second corresponds to the characterization of the thermal properties of solid materials. The latter requires an active thermal probe, from which the heat dissipation to the sample provides relevant data, to deduce some local thermophysical quantities [1,2]. We have shown that a 2-wire micro-thermocouple is able to operate in these two modes of measurement [3,4]. To complete the SThM probe, a combination of a thermocouple sensor and a quartz tuning fork (QTF) force sensor has been developed [4,5]. In the most common available systems, a resistive thermal sensor is integrated on a micro-fabricated cantilever whose deflection is optically detected to derive the contact force between the probe tip and the surface. Unlike that, the dynamic force detection principle of a QTF consists to derive an interaction force from its resonance frequency shift while the contact occurs. In this case, the embedded thermocouple tip acts as a spring, which interacts with the sample surface. If the QTF quality factor is high enough (above 5000 typically) any resonance frequency change can be interpreted as a force and used as a reliable tip-to-sample distance control loop. We first present the design of a thermal/force sensor. The experimental setup is then described and examples of topographical and thermal results in active and passive modes are depicted.

2. Thermocouple probe design and microscopy setup

Here we focus on recent advancements in the development of a SThM system based on a probe associating a micro-thermocouple with a Quartz Tuning Fork (QTF) force sensor [4,5]. "Wollaston" wires (Pt and Pt-10%Rh) are used to make the micro-thermocouple which is directly connected to QTF

IOP Conf. Series: Journal of Physics: Conf. Series **785** (2017) 012005 doi:10.1088/1742-6596/785/1/012005

electrodes using conductive adhesive. Figure 1 shows the tuning fork with its thermocouple probe. The thermocouple junction is obtained by welding two wires using a sparking technique. Among the eight available electrodes of a QTF, two electrodes are disconnected and reserved for the thermocouple connexions, and a third is used as a RTD sensor to provide the reference thermocouple temperature.

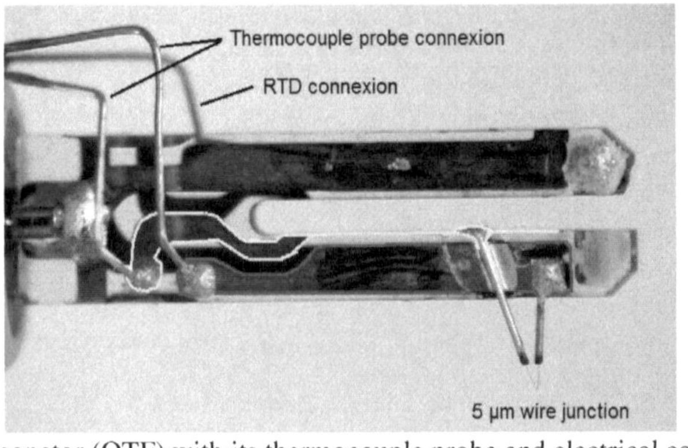

Figure 1. Quartz resonator (QTF) with its thermocouple probe and electrical connection principle.

The shape of the thermocouple junction after ion etching (FIB) is depicted in Figure 2. It clearly shows that a very sharp tip can be obtained in order to ensure a quasi-punctual contact on the sample surface and subsequently the best possible lateral resolution. Three wire diameters are available and used for probe design: 5 µm, 2.5 µm and 1.3 µm.

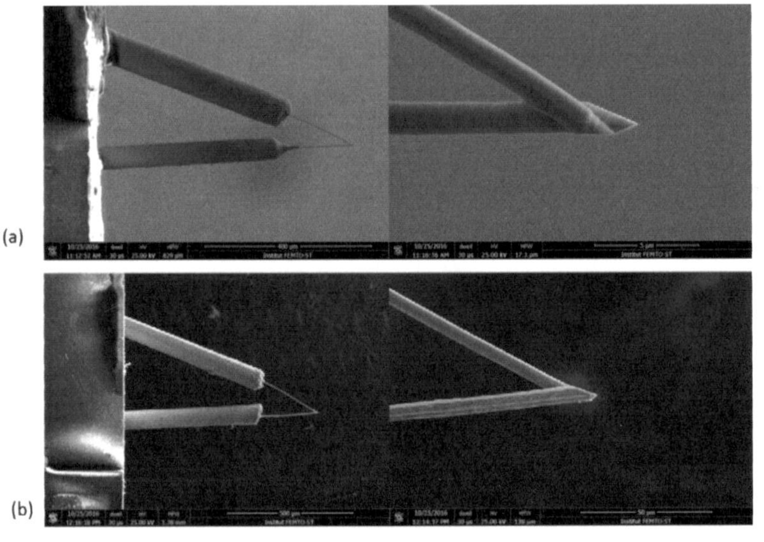

Figure 2. Thermocouple junctions: a) 1.3 µm probe and b) 5 µm probe.

The QTF is excited thermally by a laser whose magnitude is modulated at its main resonance frequency as shown in Figure 3. The photo-thermal excitation allows us to obtain a very high quality factor (10000 in air), eliminates the anti-resonance signal and parasitic capacitance effects.

The new system integrates horizontal X-Y piezo-actuators and stepper motors on which samples are placed in order to provide high resolution and large path length respectively. The probe holder is fixed along the Z axis which is associated with an independent linear motor stage and a high resolution piezo-actuator as shown in Figure 3.

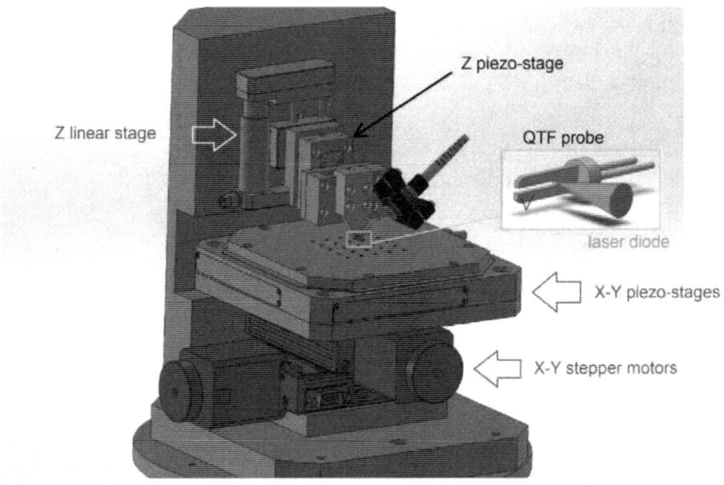

Figure 3. 3D view of the second generation FEMTO-ST SThM system.

The interaction force between the probe and the surface modifies the QTF's resonance frequency. This can be detected by measuring its phase-shift using a lock-in amplifier (open-loop) or frequency shift (close-loop) using a phase locked loop.

Figure 4 depicts the different signal responses of the thermocouple in contact with a silicon surface. When the tip-surface contact occurs, the QTF resonance frequency is shifted (figure 4a). Because the thermocouple probe is considered as a spring [5], the contact force between probe and surface can be deduced from the QTF frequency shift value and the tip's displacement as described in the Figure 4c.

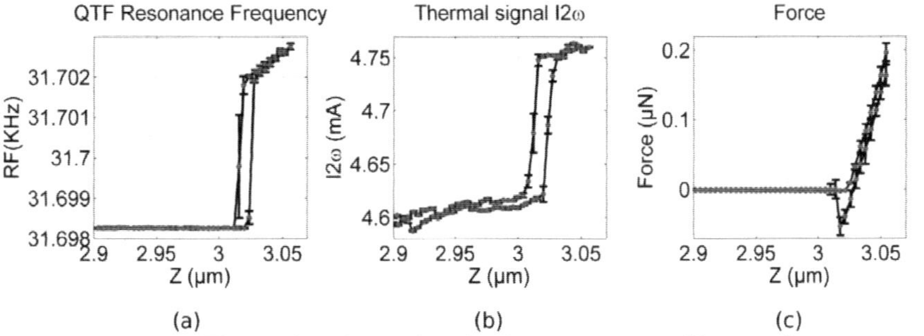

Figure 4: Effects of probe tip contact on a silicon surface.

Figure 4b corresponds to the thermal response of the junction due to the Seebeck effect at the double frequency of the AC heating current inside the probe wires [3]. It corresponds to the current value required to ensure a constant tip temperature (active 2ω mode). In these graphs, the error bars have been obtained by recording 30 consecutive values for every 2 nm step.

3. Active and passive mode measurements

To demonstrate the performance of our system, an electrothermal low power consumption device (micro-hotplate) which is composed of a thin insulating membrane of silicon nitride (500 nm thick) and a 150 nm suspended platinum heater, is used. Figure 5 presents the sample side view with the probe setup (a) and a top view of the platinum heater on the silicon nitride membrane (b). In passive mode, the platinum heater is supplied with a DC current. A scan is performed on the membrane surface (1mm x 1 mm) to extract a temperature map of the whole device surface. Absolute temperature is obtained using the S type thermocouple conversion law using the reference given by the embedded RTD sensor.

In active mode (2ω, 3ω method), the heater is not supplied whereas the thermocouple is heated with an AC current at ω frequency. When the contact occurs, the probe's temperature is affected by the thermal conductance with the sample surface, notably related to its thermal conductivity and topography (roughness). If the current is constant, one can extract both the variation of 2ω and 3ω components. The conversion of the thermoelectric voltage is based on the knowledge of the Seebeck coefficient at the mean junction temperature (DC value).

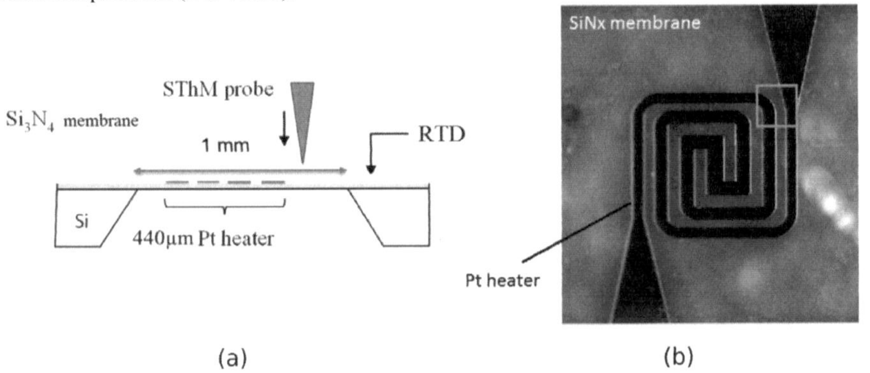

Figure 5. Side-view (a) and top-view (b) of the micro-hotplate.

Passive mode (for the probe)
During the contact, the probe temperature is obtained directly by measuring the Seebeck's voltage. The calibration methodology was already presented using standard hotplates or an optimized device with an integrated temperature sensor [6]. The real surface temperature is calculated thanks to the calibration processes. The temperature map of the device and the topography obtained when the heater operates have shown the sensitivity of a QTF in terms of force detection and the high stability of the thermocouple for temperature measurement. The system is able to reach 0.1°C of temperature resolution and works in a very wide range of temperature (from ambient to 800°C).

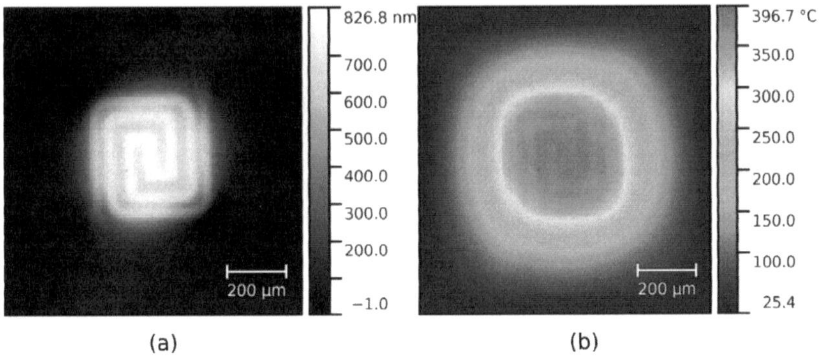

Figure 6. Topography (a) and surface temperature map (b) of active micro-hotplate.

Active mode
To illustrate the possibility to obtain the thermal contrast of a sample surface with our system, a smaller scan area (100×100 μm^2) represented by the yellow rectangle in Figure 5b has been performed. In this experiment, the sample is passive. The probe tip temperature is extracted at the double frequency of the exciting current allowing an increase in its temperature by Joule effect. Then, it is possible to measure two thermal components: the 2ω thermoelectric voltage that corresponds to the junction temperature, called 2ω method and the 3ω voltage derived from the resistive effect and the mean probe temperature

(3ω method). Since the chosen frequency of this current is low, the measured 2f temperature is identical to the DC temperature variation. Figure 7b and c show the thermal contrast obtained from 2ω and 3ω methods respectively. The tip temperature generally decreases when it approaches the platinum step. At the boundary of two materials, the tip temperature is irregularly high or low due to the topographical artifacts. A stronger cooling is measured in the middle of platinum step, far away from silicon nitride membrane due to maximum heat diffusion into the sample.

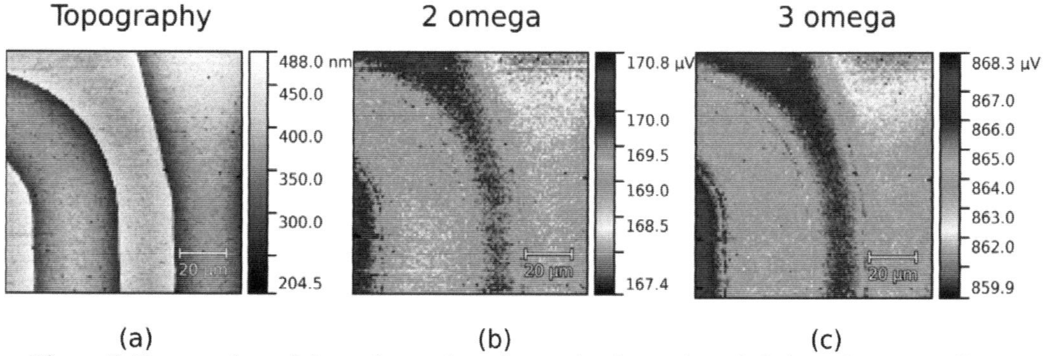

Figure 7. Topography and thermal maps in active mode of a portion of platinum heater on silicon nitride membrane with a 5μm thermocouple probe. The sample is passive.

4. Conclusions

The combination of a microwire thermocouple with a QTF has been demonstrated polyvalent. The probe can operate correctly in active mode as well in passive mode. The probe voltage is a highly stable in both modes and the available temperature range is higher than any other SThM system. Our system can also perform scans over area of millimeters square for big samples or smaller areas with resolution down to sub-nanometer scale. The system can operate at different pressures from ambient to 10^{-5} bar. Further measurements and calibration procedures under air and under vacuum conditions are still to be performed in order to obtain quantitative results. The possibility to vary the wire diameter of the thermocouple probe will be also helpful for a better understanding of the thermal spatial resolution significance.

Acknowledgments

The authors would like to acknowledge the developer group of Gwyddion for SPM data visualization software. They would also like to thank the European Union Seventh Framework Program FP7-NMP-2013-LARGE-7 under GA N°604668 Project "QUANTIHEAT" for funding this work.

References

[1] S. Gomès and S. Lefèvre, in *Advanced Techniques and Applications on Scanning Probe Microscopy*, F.H. Lei, Ed. Research Signpost, 2008, pp.157–195.

[2] B. Cretin, S. Gomès, N. Trannoy, and P. Vairac, in *Microscale and Nanoscale Heat Transfer*, Springer 2007, pp.181–238.

[3] L. Thiery, E. Gavignet and B. Cretin, "Two omega method for active thermocouple microscopy", *Rev. Sci. Instrum.* 80, 034901 (2009).

[4] A. Bontempi, D. Teyssieux, J.-M. Friedt, L. Thiery, D. Hermelin, and P. Vairac, "Photo-thermal quartz tuning fork excitation for dynamic mode atomic force microscope," *Appl. Phys. Lett.* **105**, 15, 154104 (2014).

[5] A. Bontempi, T. P. Nguyen, R. Salut, L. Thiery, D. Teyssieux, and P. Vairac, "Scanning thermal microscopy based on a quartz tuning fork and a micro-thermocouple in active mode (2ω method)," *Rev. Sci. Instrum.*, vol. 87, no. 6, 2016.
[6] E. Lemaire, T. P. Nguyen, A. Bontempi, L. Thiery, D. Teyssieux, P. Vairac, H. Shea, and D. Briand, "Micromachined calibration chip with heat source and temperature sensors for Scanning Thermal Metrology (SThM)," *Procedia Eng.*, vol. 120, pp. 130–133, 2015.

Thermal expansion coefficient of graphene using molecular dynamics simulation: A comparative study on potential functions

Hamid Ghasemi[1] and Ali Rajabpour[1,2]

[1] Mechanical Engineering Department, Imam Khomeini International University, Qazvin, Iran
[2] School of Nano Science, Institute for Research in Fundamental Sciences (IPM), Iran.

E-mail: rajabpour@eng.ikiu.ac.ir

Abstract. In this paper, we studied the thermal expansion coefficient (TEC) of pristine graphene sheets (GSs) using molecular dynamics (MD) simulation. We validated our model with previous studies employing AIREBO potential function and repeated the same simulation with the optimized Tersoff potential function. We also discussed the differences of the results and the corresponding reasons: evaluating the negative TEC of graphene by measuring the C-C bond length and out-of-plane vibrations of the GS. We finally showed that the ripples and wrinkles are more represented over the GS during the simulation with the AIREBO potential function rather than the optimized Tersoff. Comparing the results of both potential functions; it is seen that the results obtained by AIREBO potential function are in better agreement with those reported by previous scholars.

1. Introduction

Graphene has exceptional physical properties[1]–[3]; high thermal conductivity[4]–[9] and high mechanical strength make it an ideal candidate in order to control heat transfer in high-speed, high-density developed electronic and thermoelectric devices. Electronic devices usually experience wide ranges of temperature. Thermal deformations (expansion or contraction) caused by temperature differences in GSs are more significant comparing to other carbon structures: carbon nanotubes and diamonds[10], [11]. An exceptional characteristic of graphene is its negative TEC which has attracted many attentions. Therefore, studying thermal properties of graphene (especially TEC due to its interesting behaviour) over various temperatures is critical in optimizing the designs of novel graphene-based electronic devices. Many scientists have reported a negative TEC in the temperature range of 200–400 K[12]–[18]. Several authors assume that the negative TEC of GSs is due to competitions of two factors: bond length stretching and in-plane contraction due to existence of ripples and wrinkles over the GSs[15]–[17]. The prior causes a positive TEC and the later leads to an opposite effect. Although there are many experimental reports and literatures for calculating the TEC of GSs, the use of computer simulations is much less usual.

Mounet[15] used density-functional theory (DFT) and Zacharchenko[18], [19] used atomistic Monte-Carlo simulations to find TEC. Islam et al.[20] examined the effects of vacancies on the TEC of GSs. They made a comparison and found that the TEC differs less than 9% between the armchair and zigzag directions; two vacant atoms (about 1.56% by volume) increase the negative TEC nearly 40%.

Newton et al.[21] studied the thermal expansion behaviour of holes in graphene nanomeshes (GNMs). Using MD simulations, they exhibited that the GNMs behave in an opposite way: while the entire structure contracts (expands), the nano-holes expand (contract) under thermal excitation. Hu et al. [21] used lattice dynamics analysis to demonstrate that at low temperatures, ripples and negative thermal expansion, are both intrinsic characteristics of any 2D crystals with a honeycomb structure (or any monatomic 2D crystals). Based on their calculations, using self-consistent phonon theory, the negative TEC is largely affected by the "vibrational elongation" or in other words by large out-of-plane fluctuations at finite temperatures.
In this paper, we utilized MD simulations to show the effect of two popular potential functions: AIREBO and optimized Tersoff to obtain TEC and some other parameters (bond length stretching and out-of-plane fluctuations) of a GS. We explained the differences of the results and discussed the reasons. We also showed that the AIREBO potential function considers more parameters than the optimized Tersoff and maybe this is the reason of its accurate prediction of TEC.

2. Molecular dynamics simulation

MD simulations are implemented in this paper to study TEC's dependency of a GS to the temperature. A molecular dynamics simulation is based on the explanation of a potential function, or a description of the relations by which the particles in the simulation will interact. Here, we utilize two potential functions mostly used for carbon based materials: nanotubes, graphene, and graphite. LAMMPS package[22], [23] was hired in this study, using the two mentioned potential functions: optimized Tersoff and AIREBO.

2.1. Interatomic molecular dynamics potentials

2.1.1. Optimized Tersoff. The optimized Tersoff potential function predicts the interatomic interactions as follows:

$$V_{ij} = f_{ij}^{C}\left(a_{ij}f_{ij}^{R} - b_{ij}f_{ij}^{A}\right) \tag{1a}$$
$$f_{ij}^{R} = Ae^{-\lambda_1 r_{ij}} \tag{1b}$$
$$f_{ij}^{A} = Be^{-\lambda_2 r_{ij}} \tag{1c}$$

where, r_{ij} is the distance between atoms i and j and f_{ij}^{C} shows the truncation function that limits nearest neighbors atomic interactions; f_{ij}^{R} and f_{ij}^{A} represent mutual attractive and repulsive terms respectively, and a_{ij} relates to the range-limiting repulsive potential term which is typically equal to one as we do here. b_{ij} is the term depends on the bond angle and local coordination of atoms around atom i. The corresponding parameters for the optimized Tersoff-2010 potential function are represented by Lindsay et al[24].

2.1.2. AIREBO. The AIREBO model originated from a REBO one with extra LJ and torsional terms

$$V_{ij} = V_{ij}^{REBO} + V_{ij}^{LJ} + \sum_{k\neq i,j}\sum_{l\neq i,j,k} V_{ij}^{TORSION} \tag{2a}$$
$$V_{ij}^{LJ} = 4\varepsilon\left[\left(\frac{\sigma}{r_{ij}}\right)^{12} - \left(\frac{\sigma}{r_{ij}}\right)^{6}\right], r_{ij} < r_c \tag{2b}$$

where, ε is the energy parameter, σ is the distance one and r_c is the cutoff distance. All mentioned parameters are represented by Stuart et al[25].

2.2. Simulation details

In solids, thermal expansion dependency to pressure is ignored and is simply given by the following relation

$$\alpha = \frac{1}{L}\frac{dL}{dT} \tag{3}$$

where, L is the equilibrated length of the graphene as a function of equilibrium temperature T; it is recorded for each value of temperature from 50K to 1000K along x and y directions separately. We did the simulation within the NPT ensemble imposing periodic boundary conditions along x and y directions. The Newton's equations of motion were integrated with the velocity-Verlet integrator with the time step of 0.5 fs, for a total of 3 900 000 time steps (or 1.95 ns) to ensure full equilibration of the structure. Then, we made an average on length: $L = \frac{L_x+L_y}{2}$ and plotted the length values versus the temperature. Using try and error to reach the best fitted polynomial, we fitted a fourth order polynomial to the length values. According to the above equation, we obtained the TEC in each temperature. Each simulation was repeated four times in order to report the statistical uncertainties.

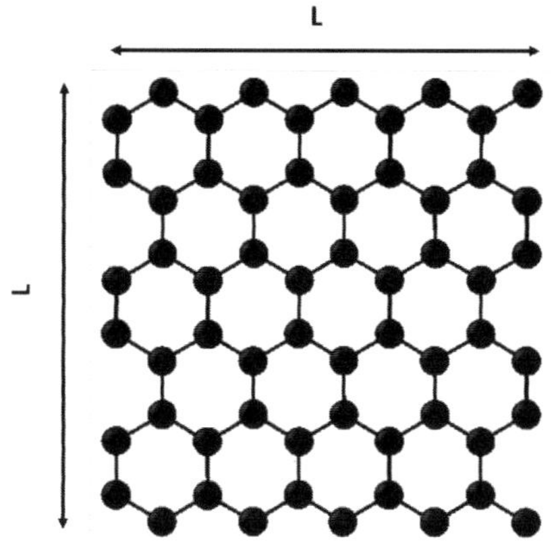

Figure 1. The schematic view of a GS.

We studied three GS lengths; 1nm, 5nm and 10nm and calculated the TEC of each size twice, once running the simulation for AIREBO potential function and then for the optimized Tersoff. Obtaining TEC values for each size, we were able to see how the size of the GS affects the TEC.
Looking for the reason of the odd behaviour – negative values in some temperatures- of the GS, we studied changing of bond length of the GS with increasing the temperature. Another important factor for this behaviour is existence of intrinsic wrinkles and ripples over the GS surface. So, we analysed how much theses ripples and wrinkles affect TEC. Finally, these two factors were compared for both optimized Tersoff and AIREBO potential functions. Then, we discussed which potential function would be a better choice in order to calculate the TEC.

3. Results and discussion

In this section, we represent our results and figures. Then, we discuss in details why the GS has this interesting behaviour by increasing the temperature. Figure 2 shows how the length of the 5 nm GS changes with increasing the temperature. It is seen that, GS has an interesting behaviour by increasing the temperature. For the simulation running with AIREBO, the length of the GS decreases as long as the temperature reaches to nearly 600 K. In another simulation with optimized Tersoff, the GS's odd behaviour does not last long; just nearly the temperature of 200K, the length starts to get increased.

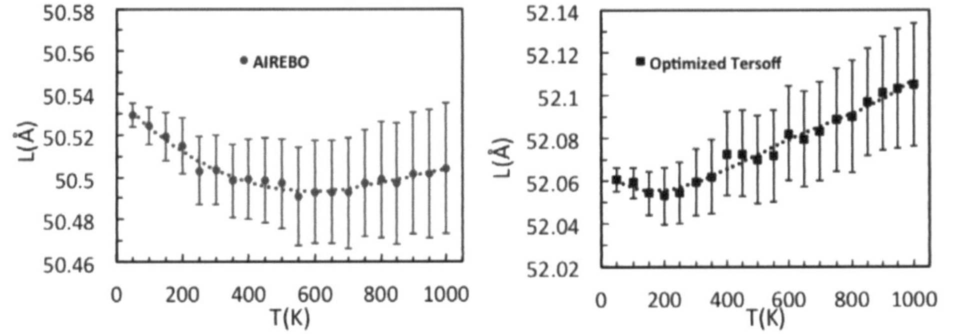

Figure 2. The length of the 5 nm GS with increasing the temperature using AIREBO potential function (Left) and optimized Tersoff potential function (Right).

The TEC results are shown in figure 3; it is seen than in AIREBO potential function, the TEC has a negative value under 600K and then changes to positive above that temperature.

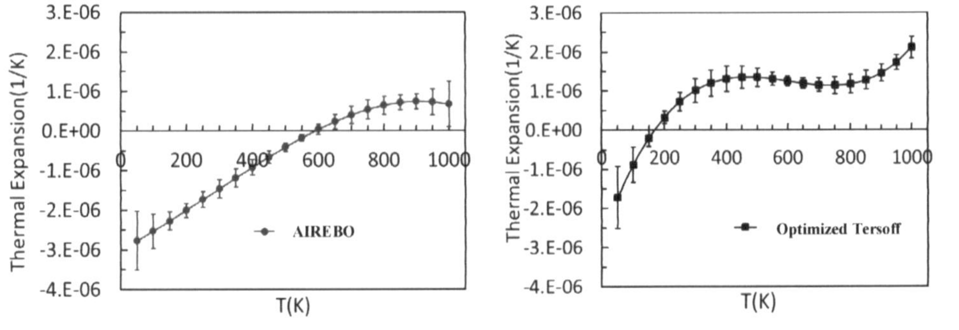

Figure 3. The TEC of the 5 nm GS with increasing the temperature using AIREBO potential function (Left) and optimized Tersoff potential function (Right).

Actually, what we have measured and called as out-of-plane fluctuations and bond length have been shown in figure 4.

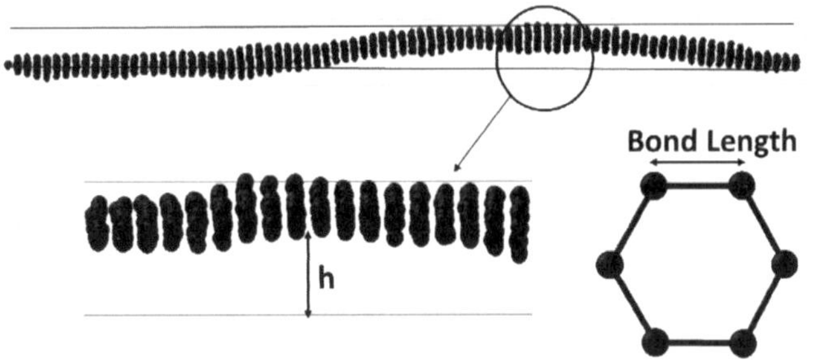

Figure 4. The schematic amplitude and bond length of a GS.

In order to calculate the amplitude (out-of-plane vibrations) of the graphene, we used a simple definition called, the standard deviation of height component of atoms.

IOP Conf. Series: Journal of Physics: Conf. Series **785** (2017) 012006 doi:10.1088/1742-6596/785/1/012006

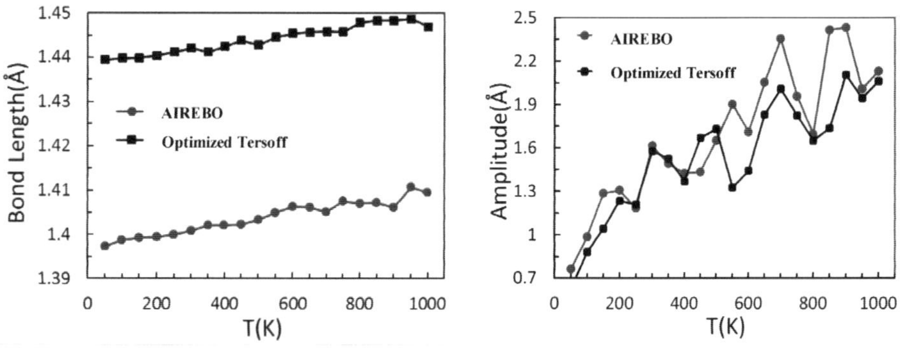

Figure 5. a) The bond length changes of carbon atoms with both AIREBO and optimized Tersoff potential functions. b) Amplitude fluctuations of carbon atoms with both AIREBO and optimized Tersoff potential functions of the 5nm sample.

The bond length changes of carbon atoms are shown in figure 5a. It is seen that, in both interaction functions, the bond length increases by increasing the temperature. So, there might be another reason for odd negative behaviour of TEC. It also means that we can have a regular (positive) behaviour for the TEC of the GS with imposing the sheet to an initial strain in order to avoid intrinsic ripples and wrinkles.

Figure 5b represents the amplitude fluctuations of the GS in different temperatures with both AIREBO and optimized Tersoff potential functions. It is clearly seen that, the overall trend of the figure is additive. Figure 6 sows the thermal expansion of graphene for various lengths. It is seen that, by increasing the temperature both potential functions predict a converging behaviour

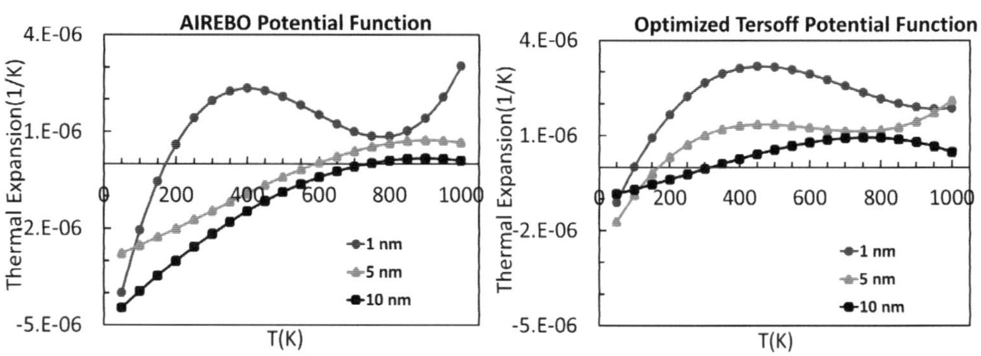

Figure 6. The effect of length on TEC of GSs.

Now, we are going to evaluate if there is an agreement between our results with those predicted by other scientists. In figure 7, we showed that our 10 nm sample which its atoms have interacted with AIREBO potential function have a better agreement with the DFT results reported by Mounet et al.[15] at temperatures higher than 200K. However, the results of classical molecular dynamics should not be compared to DFT results at low temperatures wherein quantum effects are important.

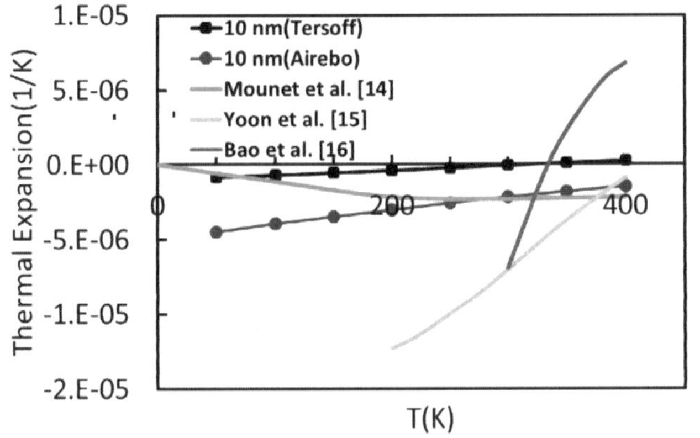

Figure 7. Comparison the results of the current study with some previous reports.

4. Conclusion

In this paper, we tried to have a review on effective reasons on interesting behaviour of TEC of pristine sheets using molecular dynamics simulations employing two interatomic potential functions: AIREBO and optimized Tersoff; we observed that by increasing the length of the graphene, the TEC shows better converging performance. On the other side, we showed how selecting a proper potential function in predicting the TEC could be important; AIREBO has a better behaviour in comparison to optimized Tersoff. It was found that the results of a 10 nm long sheet has a good agreement with those predicted by previous scholars.

5. References

[1] K. S. A. Novoselov *et al.*, "Two-dimensional gas of massless Dirac fermions in graphene," *Nature*, vol. 438, no. 7065, pp. 197–200, 2005.

[2] N. M. R. Peres, "Colloquium: The transport properties of graphene: An introduction," *Rev. Mod. Phys.*, vol. 82, no. 3, p. 2673, 2010.

[3] Y. Zhang, Y.-W. Tan, H. L. Stormer, and P. Kim, "Experimental observation of the quantum Hall effect and Berry's phase in graphene," *Nature*, vol. 438, no. 7065, pp. 201–204, 2005.

[4] W. Cai *et al.*, "Thermal transport in suspended and supported monolayer graphene grown by chemical vapor deposition," *Nano Lett.*, vol. 10, no. 5, pp. 1645–1651, 2010.

[5] S. Chen *et al.*, "Raman measurements of thermal transport in suspended monolayer graphene of variable sizes in vacuum and gaseous environments," *ACS Nano*, vol. 5, no. 1, pp. 321–328, 2010.

[6] A. a Balandin *et al.*, "Superior Thermal Conductivity of Single-Layer Graphene 2008," *Nano Lett.*, vol. 8, pp. 902–907, 2008.

[7] C. Faugeras, B. Faugeras, M. Orlita, M. Potemski, R. R. Nair, and A. K. Geim, "Thermal conductivity of graphene in corbino membrane geometry," *ACS Nano*, vol. 4, no. 4, pp. 1889–1892, 2010.

[8] J.-U. Lee, D. Yoon, H. Kim, S. W. Lee, and H. Cheong, "Thermal conductivity of suspended pristine graphene measured by Raman spectroscopy," *Phys. Rev. B*, vol. 83, no. 8, p. 81419, 2011.

[9] J. H. Seol *et al.*, "Two-dimensional phonon transport in supported graphene," *Science (80-.).*, vol. 328, no. 5975, pp. 213–216, 2010.

[10] P. K. Schelling and P. Keblinski, "Thermal expansion of carbon structures," *Phys. Rev. B*, vol. 68, no. 3, p. 35425, 2003.

[11] M. U. Kahaly and U. V Waghmare, "Effect of curvature on structures and vibrations of zigzag

carbon nanotubes: A first-principles study," *Bull. Mater. Sci.*, vol. 31, no. 3, pp. 335–341, 2008.

[12] E. G. Steward, B. P. Cook, and E. A. Kellett, "Dependence on temperature of the interlayer spacing in carbons of different graphitic perfection," 1960.

[13] A. Fasolino, J. H. Los, and M. I. Katsnelson, "Intrinsic ripples in graphene," *Nat. Mater.*, vol. 6, no. 11, pp. 858–861, 2007.

[14] H. Jiang, B. Liu, Y. Huang, and K. C. Hwang, "Thermal expansion of single wall carbon nanotubes," *J. Eng. Mater. Technol.*, vol. 126, no. 3, pp. 265–270, 2004.

[15] N. Mounet and N. Marzari, "First-principles determination of the structural, vibrational and thermodynamic properties of diamond, graphite, and derivatives," *Phys. Rev. B*, vol. 71, no. 20, p. 205214, 2005.

[16] D. Yoon, Y.-W. Son, and H. Cheong, "Strain-dependent splitting of the double-resonance Raman scattering band in graphene," *Phys. Rev. Lett.*, vol. 106, no. 15, p. 155502, 2011.

[17] W. Bao *et al.*, "Controlled ripple texturing of suspended graphene and ultrathin graphite membranes," *Nat. Nanotechnol.*, vol. 4, no. 9, pp. 562–566, 2009.

[18] K. V Zakharchenko, J. H. Los, M. I. Katsnelson, and A. Fasolino, "Atomistic simulations of structural and thermodynamic properties of bilayer graphene," *Phys. Rev. B*, vol. 81, no. 23, p. 235439, 2010.

[19] K. V Zakharchenko, M. I. Katsnelson, and A. Fasolino, "Finite temperature lattice properties of graphene beyond the quasiharmonic approximation," *Phys. Rev. Lett.*, vol. 102, no. 4, p. 46808, 2009.

[20] M. Z. Islam, M. Mahboob, L. R. Lowe, and E. S. Bechtel, "Characterization of the thermal expansion properties of graphene using molecular dynamics simulations," *J. Phys. D. Appl. Phys.*, vol. 46, no. 43, p. 435302, 2013.

[21] N. C. B. Mostério and A. F. Fonseca, "Thermal expansion behavior of holes in graphene nanomeshes," *Phys. Rev. B*, vol. 89, no. 19, p. 195437, 2014.

[22] L. Atomic and M. M. P. Simulator, "LAMMPS." 2013.

[23] S. Plimpton, "Fast parallel algorithms for short-range molecular dynamics," *J. Comput. Phys.*, vol. 117, no. 1, pp. 1–19, 1995.

[24] L. Lindsay and D. A. Broido, "Optimized Tersoff and Brenner empirical potential parameters for lattice dynamics and phonon thermal transport in carbon nanotubes and graphene," *Phys. Rev. B*, vol. 81, no. 20, p. 205441, 2010.

[25] S. J. Stuart, A. B. Tutein, and J. A. Harrison, "A reactive potential for hydrocarbons with intermolecular interactions," *J. Chem. Phys.*, vol. 112, no. 14, pp. 6472–6486, 2000.

Experimental study of multi-scale heat transfer characteristics at pool boiling

V Serdyukov, A Surtaev

Kutateladze Institute of Thermophysics, Novosibirsk, 1 Acad. Lavrentiev Ave., Russia

Abstract. This study presents the results of the experimental investigation of local and integral characteristics of heat transfer at liquid pool boiling. Saturated ethanol and water were used as the working fluids. Thin, resistively heated indium-tin oxide films deposited onto the sapphire substrates were used as the heaters. The synchronized measurements of the heater surface temperature field and dynamics of vapor bubbles were performed by high-speed infrared thermography with the frame rate of 1000 fps and resolution of up to 0.13 μm/px and high-speed video recording. In this paper new data on major local boiling characteristics, such as nucleation site density, dynamics of vapor bubbles, temporal characteristics and nucleation frequency at different heat fluxes and superheating and their comparison with correlations are presented.

1. Introduction

Nucleate boiling is one of the most effective heat transfer modes, so it is widely used in various technological applications: thermal and nuclear power engineering, chemical technology, thermal stabilization of intense heated equipment, etc. Despite a large number of papers published since the middle of the last century, devoted to the study of different aspects of boiling liquid, many questions remain open, and there is no generally accepted boiling theory [1]. This is associated with the fact that boiling is very complicated process by the need to consider phenomena that occur over multiple length scales, ranging from centimeters down to a few nanometers. In particular, the importance of evaporation from a liquid microlayer with a thickness of several microns, developing underneath a bubble during its initial growth, is still debated in boiling society.

Also in the literature there are almost no experimental data on the boiling local characteristics, such as nucleation site density, temporal characteristics of ebullition cycle in a wide range of heat flux density for liquids with different properties. At the same time information of these important local characteristics is very necessary for understanding the physics of boiling process and basic mechanisms of boiling heat transfer, since there is a relation between local characteristics and integral heat transfer. For example, such relation is the basis of mechanistic approach to the description of heat transfer at boiling [2], known as the RPI model. Therefore, understanding of vapor bubbles growth and departure mechanisms and possibility of qualitative and quantitative description of temporal characteristics and nucleation site density are very important in general to create a boiling theory.

The aim of the present work was to investigate the local and integral characteristics of nucleate pool boiling of various liquids with the use of modern experimental techniques, such as high-speed video recording and infrared thermography.

2. Experimental setup and methods

Experiments were carried out at pool boiling of saturated ethanol and deionized water at the atmospheric pressure. The detailed description of experimental setup and measurement techniques are

presented in [3]. A thin (1 μm) conductive indium tin oxide (ITO) with exposed area of 30 × 20 mm^2, vacuum deposited onto the 400 μm thick sapphire substrate was used as the heater. The main advantage of using the ITO film as a heater is its property to be opaque in the mid-IR spectrum (3-5 μm) and transparent for visible spectrum (380-750 nm). When it was deposited onto the thin sapphire substrates, transparent to both IR and visible spectrum, it became possible to investigate the unsteady temperature field of ITO film by infrared recording and record the dynamics of vapor bubbles and contact line with the use of high-speed video camera.
Measurements of the surface temperature field were performed by high speed infrared thermographic recording (IR) with the use of FLIR Titanium HD 570M camera (spectral range 3.7-4.8 μm). Thermographic recording was made with the frame rate of 1000 fps at resolution of up to 13 μm/px. High-speed visualization (HSV) was performed using the Vision Research Phantom v. 7.0 camera with 2000 fps and resolution of 640 × 480 (1 px ≈ 0.07 mm). A quick-response incandescent lamp located under the work area was used to synchronize IR and HSV recordings in time.

3. Results and discussion

3.1. Integral heat transfer and nucleation site density

Boiling curves of saturated ethanol and deionized water were obtained in the study with the use of the data of infrared thermography. The analysis has shown that results are in good agreement with experimental data of other researchers obtained at boiling on smooth surfaces (ITO, silicon, polished copper) [4-6]. It is well known [4, 7, 8] that the appearance of vapor bubble on the heating surface during nucleate boiling results in a decrease of local temperature. On the infrared images, these regions look like the darker spots (cold spots). Counting of cold spots number allows us to analyze active nucleation site density (*NSD*) at boiling in the wide range of heat fluxes up to crisis phenomena development. For this, the heater temperature fields obtained in the present study were averaged over time. Fig. 1a shows the IR image of the heater surface at ethanol pool boiling. The number of sites was automatically determined using the algorithm implemented in MatLab software. The processed image is also shown in Fig.1a, and permanent nucleation sites are labeled with markers. The error in measuring the number of sites with the use of this algorithm was less than 10%.
Nowadays there are several approaches to describe the value of nucleation site density at boiling. One of the first physically-based approaches was proposed by Labuntsov [9], who suggested that the value of nucleation site density is proportional to the number of cavities per surface, which have a size similar to the minimum critical diameter of the vapor bubbles in metastable liquid. Based on dimension analysis and taking a simple law for the size distribution, the following relationship for calculating the *NSD* value has been proposed:

$$NSD = C_0 \cdot \left(\frac{\rho_v h_{fg} \Delta T}{2\sigma T_{sat}} \right)^2 . \tag{1}$$

On the other hand, experimental data presented in the literature show that the *NSD* value depends not only on surface morphology, but also on its thermophysical and wetting properties.
In [10], the attempt to take into account the influence of thermophysical properties and roughness of heating surface on the *NSD* value at boiling was made. For analyses the authors have used experimental data obtained at pool boiling of saturated water, carbon tetrachloride, hexane and acetone on stainless steel and aluminum surfaces with different roughness. As the result of data generalization, the following dependence was proposed, where γ and Θ are the surface–liquid interaction parameter and dimensionless surface roughness parameter, respectively:

$$NSD = C_1 \Pr^{1.63} \left(\frac{1}{\gamma} \right) \Theta^{-0.4} \Delta T^3 . \tag{2}$$

As it can be seen from Fig. 1b, dependence (1) does not describe the behavior of experimental data on the *NSD* value, especially at high superheating. On the other hand, calculation using (2) with an empirical constant C_l = 92 and heating surface roughness R_a = 3 nm obtained from the morphological analysis shows good agreement with experimental data in almost whole range of superheating.

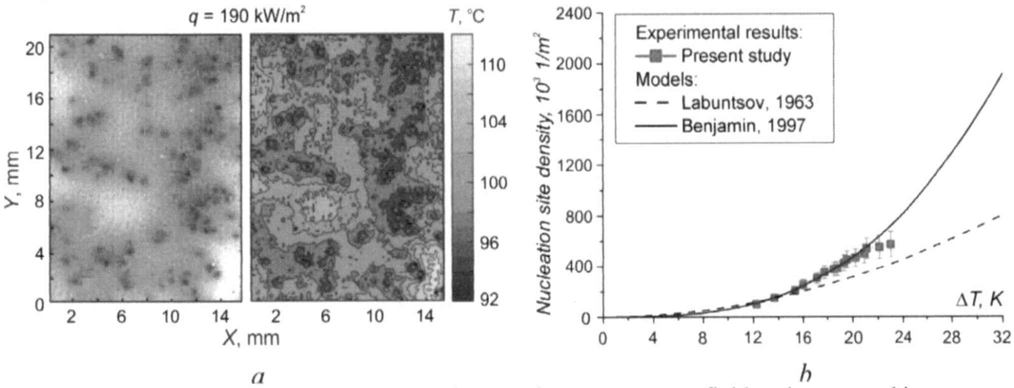

Figure 1. a) Time-averaged IR frame of the surface temperature field and processed image; b) nucleation site density at ethanol pool boiling.

It should be noted that expression (2) is semi-empirical and contains adjustable parameter C_l. Also, this model does not take into account the influence of the wetting properties of surface on nucleation site density at boiling. To create a general model for describing the nucleation site density at liquid boiling, it is necessary to provide further complex investigations.

3.2. Vapor bubble dynamics

Vapor bubble dynamics at boiling of ethanol was studied with the use of high-speed video recording (Fig. 2). It is known that at vapor bubble growth in its center there is a contact area of vapor and heating surface, the so-called "dry spot", whose boundary is the triple contact line (solid wall, liquid and vapor). It can be seen from Fig. 2 that obtained data make it possible to study both vapor bubble diameter (D_o) and dynamics of contact line in its base (D_i). The analysis of experimental data showed that the evolution cycle of a bubble can be divided into two stages. The first stage corresponds to the bubble growth. At this stage, the growth of the outer diameter of bubble is observed according to model [11]. The second stage corresponds to bubble departure from the heater. Here, a sharp decrease of the contact line area occurs with dry spot rewetting. At the departure stage, the outer bubble diameter remains constant.

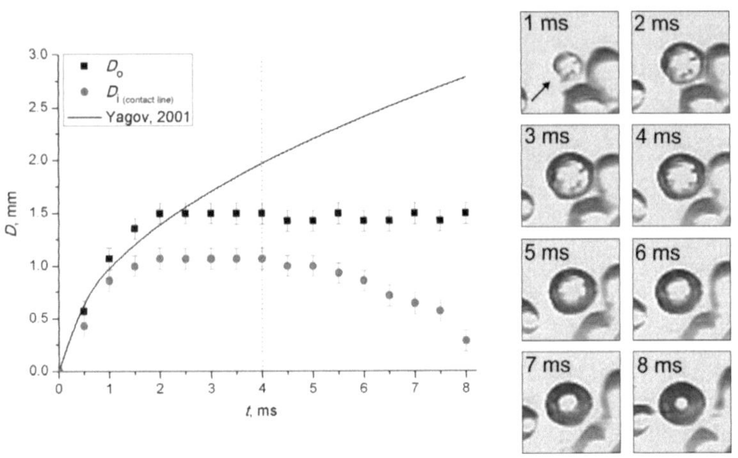

Figure 2. Dynamics of vapor bubble outer diameter D_o and contact line D_i at ethanol pool boiling (q =50 kW/m^2).

3.3. Temporal characteristics and nucleation frequency at boiling

High spatial and temporal resolutions of infrared thermography have allowed us to investigate evolution of the averaged heating surface temperature in the area under single vapor bubbles (ebullition cycle) at pool boiling of water (Fig. 3) and ethanol. It can be seen that bubble nucleation is followed by a rapid surface temperature drop. During bubble growth the temperature continues decreasing until approaching the maximum size of a bubble, corresponding to the bubble departure diameter. At the stage of bubble departure with a sharp decrease in the contact line area (as can be seen from Fig. 2), the surface temperature starts increasing monotonically. After reaching the threshold temperature of site activation, the vapor bubble appears again on the heater surface and the process repeats. The figure shows that it is reasonable to divide the ebullition cycle into two stages, characterized by fundamentally different heat transfer mechanisms: the stage of vapor bubble growth until approaching the departure diameter and the stage of its departure and transient liquid heating until new bubble appearance.

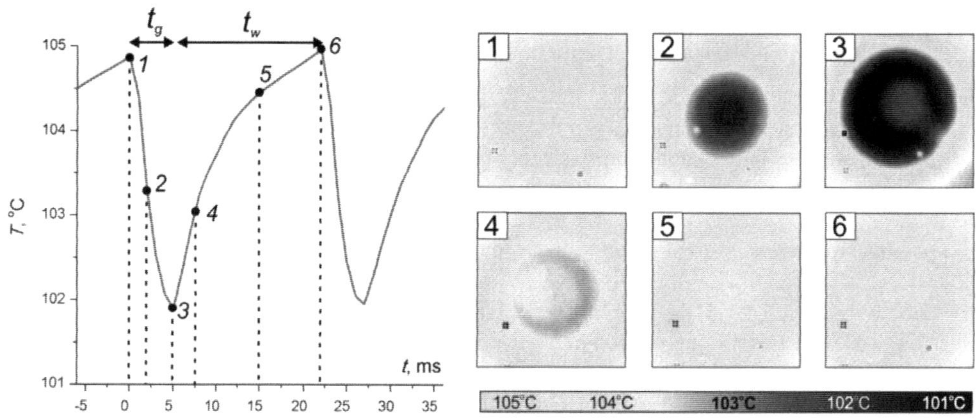

Figure 3. Ebullition cycle of single vapor bubble at boiling of water and corresponding IR frames (q =50 kW/m^2).

Dependences of bubble growth time, waiting time of new bubble appearance and nucleation frequency on heat flux density at boiling of water and ethanol were analyzed based on the obtained experimental data. It has been shown that to determine nucleation frequency, it is necessary to take into account the waiting time, which depends on the heat flux density. At water boiling in the whole range of heat fluxes (q = 0 - 450 kW/m^2) waiting time t_w exceeds the growth time of bubbles t_g. In the region of developed nucleate boiling, simple empirical relationship $t_w = 3t_g$ can be used to estimate the waiting time value [12]. For ethanol, an increase in the heat flux leads to significant reduction in the waiting time and it becomes comparable with growth time of the developed nucleate boiling region.

Nowadays, in the literature there is a relationship between nucleation frequency and bubble departure diameter in the form of $fD_{dep}^{\ n}$ = const, where n=1÷3. However, in this work it has been shown that the value of waiting time, which is necessary to take into account for nucleation frequency determination, does not depend on the value of D_{dep}. Thus, in this case the use of mentioned relationship is invalid, especially at low heat fluxes.

Also in this work the statistical analysis of the data on nucleation frequency in different areas of the heating surface at boiling of ethanol was made. It has been shown that for the given heat flux density (q = 25 kW/m^2) nucleation frequency value can vary from 20 to 100 Hz due to the different superheating thresholds of nucleation site activation in various areas of the surface. This imposes additional difficulties for comparing theoretical correlations of the nucleation frequency at boiling liquid with the experimental results.

4. Conclusion

The new experimental data on integral and local characteristics of heat transfer at pool boiling were obtained using the modern high-speed experimental techniques including infrared thermography and special design of thin-film heaters. Analysis of experimental results shows that data on nucleation site densities at pool boiling of saturated ethanol are in cubic dependence on superheating $NSD \sim \Delta T^3$ and agree with Benjamin's correlation [10]. It was shown using high-speed video recording that quite a long stage of bubble detachment occurs at ethanol boiling between the stage of vapor bubble growth and transient heating of liquid after bubble separation. At this stage, the contact line area decreases sharply, and dry spots are rewetted in short time. The temperature fields under a single vapor bubble at water and ethanol boiling were determined using IR macro recording with high spatial (~ 13 μm/pix) and temporal (1000 fps) resolution. In particular, it allowed us to estimate the periods of bubble growth and departure, time of transient heating before new bubble appearance on the surface and nucleation frequencies in dependence on the heat flux. It is shown that at theoretical description of nucleation frequency it is not correct to neglect bubble waiting time (especially in the range of low heat fluxes).

Also, the data obtained on evolution of the temperature fields under a single vapor bubble can be used in future to determine the contribution of different boiling heat transfer mechanisms (microlayer evaporation, transient heat transfer, microconvection) in the overall heat balance by numerical simulation.

5. Acknowledgments

This work was supported by the Russian Foundation for Basic Research (Grant No. 14-08-00635).

6. References

[1] Yagov V V 2014 *Heat transfer in one-phase media and at phase transformations* (in Russian) (Moscow Power Engineering Institute) p 542

[2] Podowski M 2012 Toward mechanistic modeling of boiling heat transfer *Nuclear Engineering and Technology* **8** 889–896

[3] Surtaev A S, Serdyukov V S, Moiseev M I 2016 Application of high-speed IR thermography to study boiling of liquids *Instruments and Experimental Techniques* **59(4)** 615-620

[4] Gerardi C, Buongiorno J et al 2010 Study of bubble growth in water pool boiling through synchronized, infrared thermometry and high-speed video *International Journal of Heat and Mass Transfer* **53(19)** 4185-4192

[5] Jun S, Sinha-Ray S, Yarin A L 2013 Pool boiling on nano-textured surfaces *International Journal of Heat and Mass Transfer* **62** 99-111

[6] Moita A S, Teodori E, Moreira A L N 2015 Influence of surface topography in the boiling mechanisms *International Journal of Heat and Fluid Flow* **52** 50-63

[7] Theofanous T G. et al 2002 The boiling crisis phenomenon: Part I: nucleation and nucleate boiling heat transfer *Experimental Thermal and Fluid Science* **26(6)** 775-792

[8] Golobic I et al 2009 Experimental determination of transient wall temperature distributions close to growing vapor bubbles *Heat and mass transfer* **45(7)** 857-866

[9] Labuntsov D A 1963 *Proc. of the USSR Academy of Sciences: Energy and Transport*

[10] Benjamin R J and Balakrishnan A R 1997 Nucleation site density in pool boiling of saturated pure liquids: effect of surface microroughness and surface and liquid physical properties *Experimental Thermal and Fluid Science* **15(1)** 32-42

[11] Yagov V V 2001 Bubble growth rate at pool boiling in wide range of reduced pressures *International Journal of Heat and Technology* **19(2)** 17-24

[12] Van Stralen S J D et al 1975 Bubble growth rates in nucleate boiling of water at subatmospheric pressures *International Journal of Heat and Mass Transfer* **18(5)** 655-669

IOP Conf. Series: Journal of Physics: Conf. Series **785** (2017) 012008 doi:10.1088/1742-6596/785/1/012008

Time-dependent conductive heat transfer in rarefied polyatomic gases confined between parallel plates

A Tsimpoukis[1], C Tantos, D Valougeorgis

University of Thessaly, Dept. of Mechanical Engineering, Volos 38334, GR

[1]Email: atsimpoukis@mie.uth.gr

Abstract. The transient conductive heat transfer through a rarefied gas confined between two infinite parallel plates due to a sudden jump in the temperature of one of the plates is investigated in the whole range of the Knudsen number via kinetic theory. More specifically, the time-dependent heat transfer flow is modelled by the Holway kinetic model subject to diffuse boundary conditions. The governing integro-differential equation is numerically solved using the discrete velocity method in the molecular velocity space and typical finite control volume schemes in time and physical spaces. The time evolution of the density and temperature distributions as well as of the translational and rotational heat fluxes in terms of the two parameters characterizing the heat flow, namely the Knudsen number and the imposed temperature ratio between the plates is provided. The investigation is focused on the effect of the rotational degrees of freedom and a comparison between monatomic and polyatomic gases is performed. It is found that the time needed to reach the steady-state conditions varies between monatomic and polyatomic gases. In all cases the total time to recover the stationary solution in terms of the rarefaction parameter exhibits a minimum close to the well-known Knudsen minimum.

1. Introduction

Heat transfer through stationary rarefied gases confined between parallel plates is a fundamental problem in rarefied gas dynamics. This heat transfer configuration is very common in several technological applications including vacuum solar collectors [1], micro heat exchangers and microsensors [2,3]. Heat transfer modeling in rarefied gases is commonly based on kinetic theory [4] in order to yield reliable solutions in the whole range of the Knudsen number. The case of plane and cylindrical heat conduction through monatomic gases has been examined in [5,6], while the corresponding more complex heat flow setups in polyatomic gases in [7,8] based on the Holway model and the DSMC method. All above work refers to steady-state conditions, while the corresponding work in the case of transient conditions is limited to monatomic gases [9].

The present work is devoted to the kinetic solution of time-dependent conductive heat transfer through rarefied polyatomic gases confined between parallel plates by numerically solving the time dependent Holway kinetic equation [10] subject to diffuse boundary conditions. The time evolution of all macroscopic quantities including the translational and rotational heat fluxes is obtained. The response time to reach steady-state conditions is computed. The effect of the number of rotational degrees of freedom is investigated and the differences (and similarities) compared to the corresponding monatomic gas transient heat transfer problem are pointed out.

2. Plane heat flow configuration

Consider the state of a rarefied stationary nonpolar polyatomic gas confined between two infinite parallel plates fixed at $y'=\pm H/2$, which initially are in the same temperature T_C. Then, at time $t'=0$, the temperature of the plate at $y'=-H/2$ is suddenly increased at T_H, with $T_H>T_C$ and then for $t'>0$ the temperature of the two plates are maintained constant (figure 1). As a result 1D transient heat flow from the hot towards the cold plate will commence, which will gradually grow and as $t'\to\infty$, it will asymptotically approach the corresponding steady-state heat flow. The present analysis treats only the translational and rotational energy modes ignoring the vibrational ones.

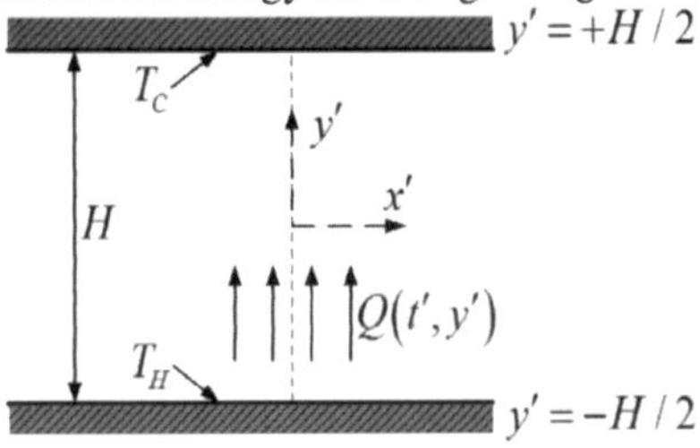

Figure 1. Heat flow configuration.

In the temperature range where the effects of vibrational degrees of freedom can be neglected, the problem may be modelled by kinetic theory for a gas of rigid rotators. When intrinsic molecular angular momenta has no preferential alignment, the gas may be described by the distribution function $f'(t',y',\upsilon,I')$, where $\upsilon=(\xi_x,\xi_y,\xi_z)$ is the molecular velocity vector and I' is the internal energy, describing the molecular internal states [11]. In polyatomic gases the internal energy is divided in the energy of the translational motion and the energy associated to the internal structure. These energies are related to the corresponding translational and rotational temperatures and heat fluxes.

Then, the macroscopic quantities of practical interest are obtained by the moments of f' as

$$n(t',y')=\int_{-\infty}^{\infty}\int_{-\infty}^{\infty}\int_{-\infty}^{\infty}\int_{0}^{\infty} f'dI'd\xi_x d\xi_y d\xi_z\,,\quad u_y(t',y')=\frac{1}{n}\int_{-\infty}^{\infty}\int_{-\infty}^{\infty}\int_{-\infty}^{\infty}\int_{0}^{\infty}\xi_y f'dI'd\xi_x d\xi_y d\xi_z \tag{1}$$

$$T_{tr}(t',y')=\frac{m}{3k_B n}\int_{-\infty}^{\infty}\int_{-\infty}^{\infty}\int_{-\infty}^{\infty}\int_{0}^{\infty}\left(\left(\xi_y-u_y\right)^2+\xi_x^2+\xi_z^2\right)f'dI'd\xi_x d\xi_y d\xi_z \tag{2}$$

$$T_{rot}(t',y')=\frac{2}{jk_B n}\int_{-\infty}^{\infty}\int_{-\infty}^{\infty}\int_{-\infty}^{\infty}\int_{0}^{\infty} f'I'dI'd\xi_x d\xi_y d\xi_z\,,\quad T=\frac{3T_{tr}+jT_{rot}}{3+j} \tag{3}$$

$$Q_{tr}(t',y')=\frac{m}{2}\int_{-\infty}^{\infty}\int_{-\infty}^{\infty}\int_{-\infty}^{\infty}\int_{0}^{\infty}\left(\left(\xi_y-u_y\right)^2+\xi_x^2+\xi_z^2\right)\left(\xi_y-u_y\right)f'dI'd\xi_x d\xi_y d\xi_z \tag{4}$$

$$Q_{rot}(t',y')=\int_{-\infty}^{\infty}\int_{-\infty}^{\infty}\int_{-\infty}^{\infty}\int_{0}^{\infty}\left(\xi_y-u_y\right)I'f'dI'd\xi_x d\xi_y d\xi_z\,,\quad Q=Q_{tr}+Q_{rot} \tag{5}$$

where n, T, Q and u_y are the number density, total (thermodynamic) temperature, total heat flux and velocity respectively, while k_B is the Boltzmann constant and m the molecular mass. The subscripts tr and rot denote the translational and rotational parts. The parameter j is the number of rotational degrees of freedom, with $j=2$ in diatomic and linear molecules and $j=3$ in all other cases ($j=0$ refers to monoatomic molecules).

The two parameters characterizing the problem are the normalized temperature difference β and the reference gas rarefaction parameter δ

$$\beta = \frac{T_H - T_C}{2T_0}, \quad \delta = \frac{P_0 H}{\mu_0 \upsilon_0} \sim \frac{1}{Kn} \tag{6}$$

where $T_0 = (T_H + T_C)/2$ is the reference temperature, μ_0 is the gas viscosity at T_0, $\upsilon_0 = \sqrt{2k_B T_0 / m}$ is the most probable molecular speed and $P_0 = n_0 k_B T_0$ is a reference pressure. The gas rarefaction parameter is proportional to the inverse Knudsen number and it is used for practical purposes.

3. Polyatomic kinetic modelling

The kinetic model introduced by Holway [10] where the collision integral consists of two components corresponding to the elastic and inelastic collisions is applied. In dimensionless form the Holway model for the present one dimensional problem is written as [8]

$$\frac{\partial \boldsymbol{\Psi}}{\partial t} + c_y \frac{\partial \boldsymbol{\Psi}}{\partial y} = \delta \rho \tau_{tr}^{1-\omega} \Pr \left(\left(1 - \frac{1}{Z}\right)(\boldsymbol{\Psi}_{tr} - \boldsymbol{\Psi}) + \frac{1}{Z}(\boldsymbol{\Psi}_{rot} - \boldsymbol{\Psi}) \right) \tag{7}$$

where $t = (t'\upsilon_0)/H$, $y = y'/H$ and $c_y = \xi_y / \upsilon_0$ are the dimensionless independent variables of time, space and $y-$component of the molecular velocity, while the unknown dependent variables are in vector form as $\boldsymbol{\Psi} = [F, G, S]^T$. Furthermore, $\boldsymbol{\Psi}_{tr} = [F_{tr}, G_{tr}, S_{tr}]^T$ and $\boldsymbol{\Psi}_{rot} = [F_{rot}, G_{rot}, S_{rot}]^T$ are the translational and rotational relaxing distributions given by

$$F_{tr} = \frac{\rho}{\pi \tau_{tr}} \exp\left[-\frac{(c_y - u_y)^2}{\tau_{tr}} \right], \quad F_{rot} = \frac{\rho}{\pi \tau} \exp\left[-\frac{(c_y - u_y)^2}{\tau} \right],$$

$$G_{tr} = \tau_{tr} F_{tr}, \quad G_{rot} = \tau F_{rot}, \quad S_{tr} = \frac{j}{2} \tau_{rot} F_{tr}, \quad S_{rot} = \frac{j}{2} \tau F_{rot} \tag{8}$$

The dimensionless macroscopic distributions of density, velocity, translational and rotational as well as total temperatures and corresponding heat fluxes are defined as

$$\rho = \frac{n}{n_0}, \quad u_y = \frac{u'_y}{\upsilon_0}, \quad \tau_{tr} = \frac{T_{tr}}{T_0}, \quad \tau_{rot} = \frac{T_{rot}}{T_0}, \quad \tau = \frac{3\tau_{tr} + j\tau_{rot}}{3 + j},$$

$$q_{rot} = \frac{Q_{rot}}{(P_0 \upsilon_0)}, \quad q_{tr} = \frac{Q_{tr}}{(P_0 \upsilon_0)}, \quad q = q_{tr} + q_{rot} \tag{9}$$

Also, in Eq. (7), Pr is the Prandtl number, while the parameter $0 \le Z^{-1} \le 1$ indicates the fraction of rotational collisions with regard to the total number of collisions. In the kinetic formulation of the problem the so-called inverse power law model (IPL) has been introduced with the viscosity varying as $(\mu / \mu_0) = (T / T_0)^{\omega}$ and ω denoting the gas collision index.

At the two plates the Maxwell boundary conditions with purely diffuse reflection are applied for the incoming distributions. Hereby, to distinguish between incident and reflected quantities the superscripts (−) and (+) are introduced. Following a well-known procedure the following reflected reduced distributions at the boundaries $y = \pm 1/2$ are deduced:

$$F^+ = \left(\rho_W / \sqrt{\pi \tau^a}\right) \exp\left(-\frac{c_y^2}{\tau^a} \right), \quad G^+ = \tau^a F^+, \quad S^+ = \left(\frac{j}{2}\right) \tau^a F^+ \tag{10}$$

where $\tau^{\alpha} = (1+\beta)$ and the parameter ρ_W is specified by the no penetration condition. The problem to be numerically solved consists of the kinetic equations (7) with the asscociated expressions (8) and (9) subject to boundary conditions (10).

The discretization of the kinetic equations in the molecular velocity space is based on the discrete velocity method. In the physical space a second order difference scheme is applied, while the macroscopic quantities are computed by Gauss-Legendre quadrature. Concerning the temporal discretization, a semi-implicit scheme is used as in [12]. The results presented in the next section have been obtained with 10^3 nodes in the physical space, 96 molecular speeds and time step $\Delta t = 10^{-4}$. The evolution with respect to time is concluded at some time t_E, where the computed density, temperature and heat flux reach up to 99% of their corresponding steady-state values.

4. Results and discussion

In figure 2 the evolution of the translational q_{tr} and rotational q_{rot} heat fluxes at the center of the gap ($y = 0$) is presented in terms of the dimensionless time t up to some certain time for typical values of $\delta = [0,1,50]$ and for $T_H/T_C = [1.1,3,10]$ corresponding to small, moderate and large temperature gradients. The relaxation times to steady-state conditions are about the same for the translational and rotational heat fluxes and they are both increased as the temperature ratio is increased. Furthermore, the time needed to reach the steady-state situation in the transition regime ($\delta = 1$) is considerably smaller than the corresponding times in the free molecular ($\delta = 0$) and hydrodynamic ($\delta = 50$) regimes.

In figure 3 the total dimensionless response time t_E is shown in terms of the rarefaction parameter δ for different values of the temperature ratio T_H/T_C for a diatomic gas, while corresponding results for a monoatomic gases are also included for comparison purposes. In all cases the required response time is reaching a minimum somewhere inside the transition regime at about $\delta = 1$. This behavior has some resemblance to the well-known Knudsen minimum. For values of $\delta > 1$, the time t_E for a polyatomic gas is larger than for the corresponding monoatomic one and for values of $\delta < 1$, the time t_E is about the same for polyatomic and monoatomic gases at small and moderate temperature ratios, while at the large temperature ratio it is larger for the monoatomic gas.

In order to obtain a more physical understanding of the required time to reach steady-state conditions and to facilitate comparisons with experiments certain dimensional times t' [s] in terms of the reference pressure P_0 [Pa] are given in figure 4 for the monatomic gases of Ar and He, the diatomic gases of H_2 and N_2 and the polyatomic gases of CO_2 and SF_6. The distance between the plates is $H = 5\,\text{mm}$ with the temperature of the cold plate set at $T_C = 273\,[\text{K}]$ and the temperature ratio $T_H/T_C = 3$. The reference pressure P_0 [Pa] is easily connected to the rarefaction parameter δ, via Eq. (6), once the viscosity and the most probable velocity of each gas is specified. All computations are based on the IPL interaction law with ω= [0.81, 0.66, 0.67, 0.74, 0.93, 0.69] for Ar, He, H_2, N_2, CO_2 and SF_6 respectively. It is noted that the dependency of the results on Z is small and therefore in all cases $Z = 5$ is introduced [8]. The minimum response times at figure 3, are also present here in the range of 1-10 [Pa] depending on the specific gas. In the case of monatomic gases as molecular mass m is increased the response time is also monotonically increased. In polyatomic gases this trend is not always valid because it also depends on the internal structure. For example the response time of N_2 is larger than of Ar even if N_2 is lighter of Ar. Of course the response times of H_2 and SF_6, which are the lightest and heaviest gases are the smallest and the largest respectively.

Closing this work it is noted that the interest of the research community in time-dependent heat transfer in rarefied gases has been lately increased mainly due to the technological applications in the field of microfluidics. It is hoped that the provided results may trigger some experimental work in time-dependent heat transfer configurations and also support the design and optimization of micro heat transfer devices.

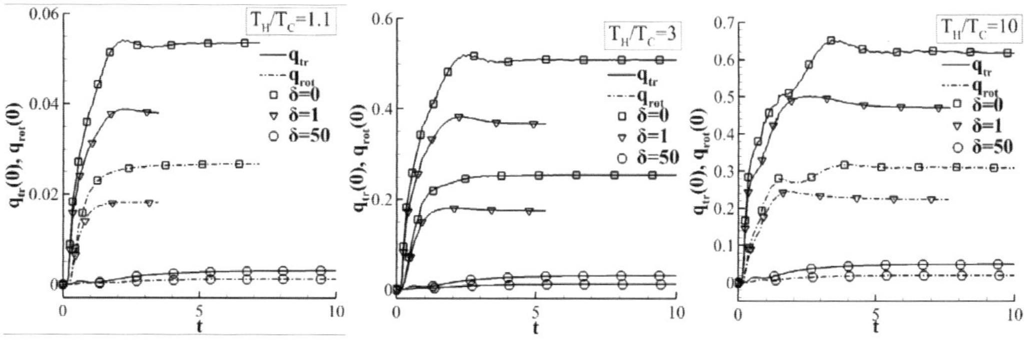

Figure 2. Time evolution of q_{tr} (solid) and rotational q_{rot} (dash dot) heat fluxes for various parameters.

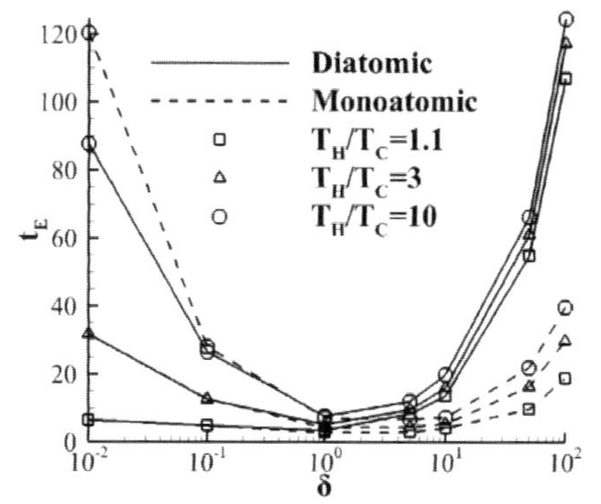

Figure 3. Response time t_E of diatomic (solid) and monoatomic gases (dashed) in terms of δ and $T_H/T_C = [1.1, 3, 10]$.

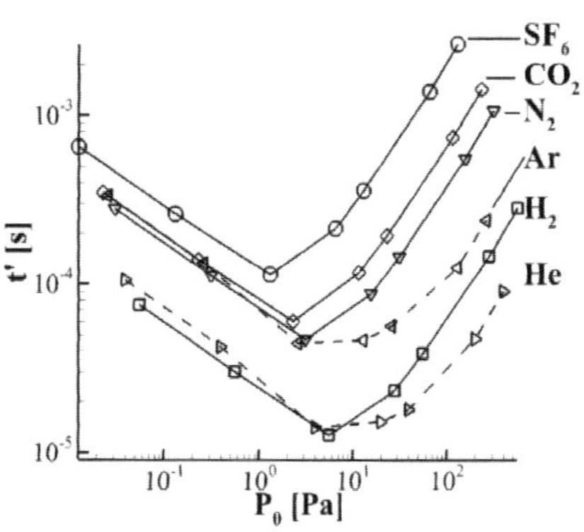

Figure 4. Response time t' [s] in terms of reference pressure P_0 with $T_H/T_C = 3$ for various polyatomic and monoatomic gases.

Acknowledgments
This work has been supported by the European Community under the contract of Association EUROfusion – Hellenic Republic 2014-2018 (Grant Agreement No 633053, WP EDU).

References

[1] O'shea S and Collins R 1992 *Int. J. Heat Mass Transfer* **35** 3431–40
[2] Yang Y, Gerken I, Brandner J, Morini G 2014 *Exp. Heat Transfer* **27** 340–59
[3] Chalabi H, Buchina O, Saraceno L, Lorenzini M, Valougeorgis D, Morini G 2012 *J. Phys: Conf. Ser.* **362**
[4] Ferziger J and Kaper H 1972 *Mathematical Theory of Transport Processes in Gases* (Amsterdam: North-Holland Publishing Company)
[5] Pantazis S and Valougeorgis D 2008 *Proc. Conf. 5th Eur. Thermal-Sciences* Eindhoven p 407
[6] Pantazis S and Valougeorgis D 2010 *Eur. J. Mech. B-Fluid* **29** 494-509
[7] Tantos C, Valougeorgis D, Pannuzzo M, Frezzotti A and Morini G 2014 *Int. J. Heat Mass Transfer* **79** 378–89
[8] Tantos C, Valougeorgis D and Frezzotti A 2015 *Int. J. Heat Mass Transfer* **88** 636-51
[9] Graur I, Ho M and Wuest M 2013 *J. Vac. Sci. Technol. A* **31** 061603
[10] Holway L 1966 *Physics of Fluids* **9** 1658-73
[11] Kuscer I 1989 *Physica A* **158** 784–800
[12] Lihnaropoulos J and Valougeorgis D 2011 *Fusion Eng. Des.* **86** 2139-42

Effect of the amorphization around spherical nano-pores on the thermal conductivity of nano-porous Silicon

Maxime Verdier, David Lacroix and Konstantinos Termentzidis

Université de Lorraine, LEMTA, CNRS-UMR7563, BP 70239, 54506 Vandœuvre Cedex, FRANCE.

E-mail: `maxime.verdier@univ-lorraine.fr`

Abstract. The thermal conductivity of nano-porous Silicon with amorphous shells around the pores is computed by Molecular Dynamics simulations. For the latter property, a systematic investigation of the porosity and the thickness of the amorphous shells has been performed. Sub-amorphous thermal conductivity is reached for systems with large porosity and amorphous shell, while a non-negligible fraction of crystalline Silicon phase is still present. The thermal conductivity of all studied systems can be controlled by a key parameter which is the ratio of crystalline/amorphous or crystalline/void interface to the volume of material.

1. Introduction

Nano-porous semiconductors have been intensively investigated during the last decade. Nano-porous Silicon has possible applications in photovoltaic devices [1] or as a non-toxic and low cost thermoelectric material. Thermal transport through nano-porous systems is often lowered because of phonon scattering on pores' surfaces. Furthermore, it has been observed that the reduction of the thermal conductivity is more pronounced when the scattering surface per unit of volume is large, *i.e.* when the porosity is distributed through a high number of small pores [2, 3, 4].

Experimentally, the elaboration of nano-pores is often accompanied by an amorphization around the pores [5]. This might be due to a local reorganization of the atoms around the holes, or to the oxidation of the surface. Amorphous and crystalline phases of semiconductors have very different thermal properties. For instance, the thermal conductivity of bulk crystalline Silicon is 150 W/m.K whereas the bulk amorphous Silicon one is only 1.5 W/m.K (divided by 100) at room temperature. So the presence of an amorphous shell around the pore should have consequences on the global thermal transport, depending upon the shell thickness. A shell of amorphous Silicon around the pore have been modeled by means of Molecular Dynamics simulations in the present work.

First, the geometrical configuration of our systems is described and technical details about the numerical method used to compute the thermal conductivity are given. Then a comparison of the thermal conductivity between nano-porous materials with and without shell around the pores is given. Finally, a key parameter, which can control the thermal conductivity of such systems, is discussed.

2. Modeling the structures

The size of the simulation box is constant for all the systems: $10a_0 \times 10a_0 \times 10a_0$ with $a_0 = 5.43$ Å the lattice constant, containing 8000 Silicon atoms. One spherical pore is created at the center of the box by erasing some atoms in the specific region. The diameter of this pore can vary from 0 to the dimensions of the box, so the maximum diameter of the pore is 5.43 nm. Periodic boundary conditions are applied on all directions. Thus, we compute the thermal conductivity of an infinite size system with spherical nano-pores periodically spread across the material. A shell of amorphous Silicon can be added around the pore with a "cut and paste technique" [6, 7, 8] . The thickness of the shell can also vary. The thermal conductivity is always computed at room temperature ($T = 300$ K).

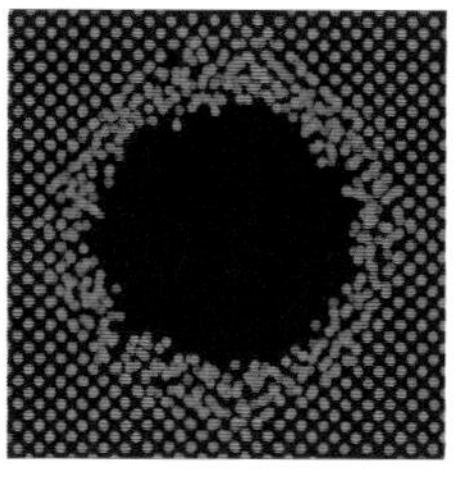

Figure 1. Visualization of a slice of a system modeled with Molecular Dynamics (VMD software [9]). Geometric parameters for this system are: $R = 3\ a_0$ and thickness of the amorphous shell $t = 0.5$ nm. Periodic boundary conditions are applied on all directions.

3. Simulation set up

The simulations are performed with LAMMPS [10]. The time integration is done with Verlet algorithm. The Stillinger-Weber three-body interatomic potential [11] with modified coefficients [12] is used to describe the interactions between the atoms. It has been chosen for its ability to provide an accurate description of thermal transport in both crystalline and amorphous phases of Silicon. The timestep is set to 0.5 fs. Before the main calculation, the system is relaxed to 300 K during 200 ps with a NVT ensemble.

Then the thermal conductivity is computed by Equilibrium Molecular Dynamics (EMD), which is based on the Green-Kubo formula:

$$k_{xy} = \frac{\mathcal{V}}{k_B T^2} \int_0^{+\infty} \langle J_x(t) J_y(0) \rangle dt \tag{1}$$

with $\mathcal{V}$ the total volume of the simulation box, T the temperature, k_B the Boltzmann constant and J the heat flux. The system evolves freely around its equilibrium state under NVT ensemble and the correlation of the fluctuations of the heat flux gives the thermal conductivity. The flux fluctuations are computed each 10 fs and the integral is sampled over 4000 values, so we obtain an estimation of the thermal conductivity every 40 ps. The total duration of a simulation is usually about 10 ns. Each computation is replicated 10 times with different initial conditions in order to reduce uncertainties. This method was validated by retrieving the thermal conductivity of bulk crystalline Silicon ($k = 170 \pm 21$ W/m.K) with different sizes of simulation box. A box of $10a_0 \times 10a_0 \times 10a_0$ was found to be large enough to guarantee that there is no size effects.

4. Effect of the amorphous shell

In this section, the thermal conductivities of nano-porous Silicon with and without amorphous shell around the pore are compared. In figure 2, the thermal conductivity is depicted as a function of porosity. When the size of the pore is increased, the porosity also increases, as the size of the simulation box is fixed. The thickness of the shell is set to 0.5 nm for all results given

IOP Conf. Series: Journal of Physics: Conf. Series **785** (2017) 012009 doi:10.1088/1742-6596/785/1/012009

in figure 2. The red line displays the thermal conductivity of bulk amorphous Silicon (about 1.5 W/m.K).

Obviously, the thermal conductivity decreases when the porosity increases, regardless of the presence or not of an amorphous shell. In figure 2, we compare the thermal conductivity of nanoporous Silicon without (black curve) and with amorphous shells (red curve). Systems with amorphous shells have systematically lower thermal conductivity than systems without shells for all the range of the studied porosity. This means that the amorphization around the pore reduces further the global thermal conductivity. This behavior is not surprising as the amorphous shell "crop" a part of the crystalline Silicon domain. This can also be seen when varying the thickness of the amorphous shell while keeping a constant radius of the pore: increasing the fraction of the amorphous phase leads to a greater reduction of thermal transport [13].

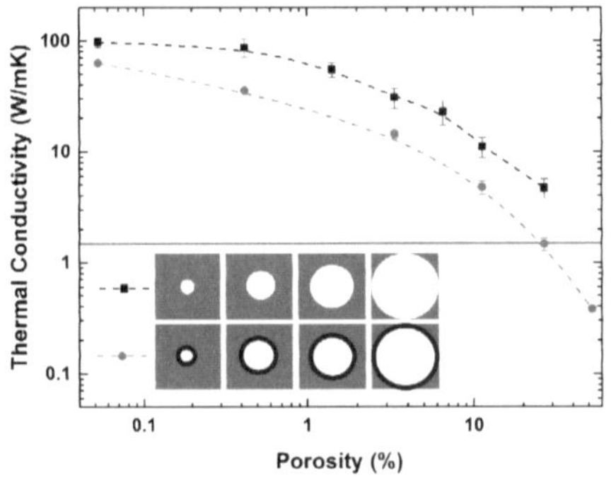

Figure 2. Thermal conductivity of nano-porous Silicon with or without amorphous shell around the pores as a function of the porosity. The radius of the pore varies inside a simulation box with fixed dimensions. The shell thickness here is set to 0.5 nm.

The thermal conductivity of some systems presented in figure 2 is extremely low, *e.g.* for pore's diameter of 4.3 nm with the amorphous shell, the porosity is greater than 25% and the thermal conductivity is about 1.5 W/m.K. So the ratio of the scattering surface over the volume is very high. For most porous systems with an amorphous shell, we can clearly see in the figure 2 that the thermal conductivity can even reach values below the amorphous limit. Furthermore, for the last point of the red curve, there is still a continuous and non-negligible fraction of crystalline Silicon in the system. When looking at the schematic representation of the system with the biggest pore in the figure 2, one can believe that there is crystalline Silicon only in the corners of the box. But the pore is spherical and this pattern displays a slice of the system in the middle, where the pore size is maximal. So, imagining the real 3-dimension system, it appears that there is still a reasonable fraction of crystalline material (about 13% of the total volume and 27% of the material volume for the last configuration) which form a continuous path between the pores and amorphous shells (a 3D array of interconnected and modulated crystalline nanowires) and may allow the transport of electrons for thermoelectric applications.

This extremely low thermal conductivity might be due to the combination of the high density of interfaces (crystalline/amorphous and amorphous/void) and the modifications of the phonon density of states inside the crystalline phase because of the proximity of amorphous material. It has been shown that phonons feel the existence of the crystalline/amorphous interface in a distance of almost 1 nm from the interface [6].

5. Thermal conductivity behavior in nanoporous Silicon, generalization

In order to understand the thermal transport behavior due to the phonons in nano-structured semiconductors with amorphous phase, we have sought a key parameter controlling the thermal conductivity for the studied systems. Usually, for homogeneous and isotropic nano-porous materials, the control parameter is taken as the ratio of the scattering surface (the surface of the pores) over the volume of material. In the case of one spherical nano-pore in a periodic and cubic simulation box, this ratio is written as:

$$\frac{S}{\mathcal{V}} = \frac{4\pi R^2}{L^3(1-\phi)} \tag{2}$$

where R is the radius of the pore, L is the dimension of the simulation box ($L = 5.43$ nm) and ϕ is the porosity.

Presently, the question is to know how we can consider the presence of an amorphous phase in a general way. In fact, in order to merge all results on the same curve, we have to consider the crystalline/amorphous interface as the main scattering surface:

$$\frac{S}{\mathcal{V}} = \frac{4\pi(R+t)^2}{L^3(1-\phi)} \tag{3}$$

with t the thickness of the shell. So the amorphous/void interface is not considered, it seems to play a minor role in thermal transport. Taking this formula to calculate the ratio $S/\mathcal{V}$ when there is an amorphous shell, results for systems with or without shell indeed gather on a same curve (fig. 3). This shows that the crystalline/amorphous interface has roughly the same effect as a crystalline/void interface on thermal transport. Another proof of this assumption is given in our precedent work [13]. Plotting same data as a function of $S/\mathcal{V}$, with S the interface between the material (crystalline or amorphous phase) and the void, the three different systems did not converge on the same curve as previously (fig. 4).

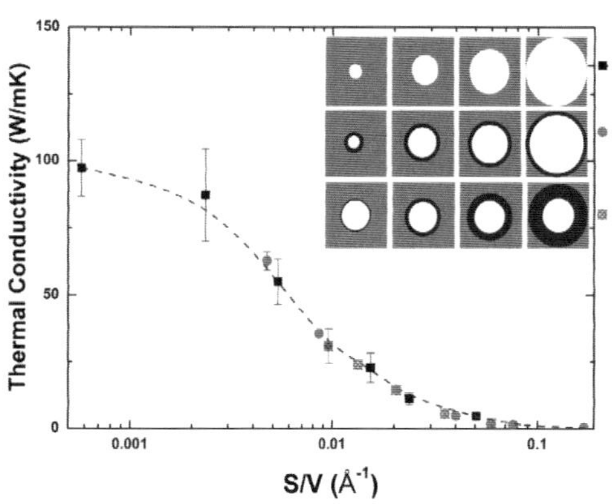

Figure 3. Generalization of the thermal conductivity of nano-porous Silicon with or without amorphous shell around the pores as a function of the ratio $S/\mathcal{V}$, with S the crystalline/amorphous or crystalline/void interface. For systems with the amorphous shell, both "constant thickness - varying radius" and "constant radius - varying thickness" configurations are displayed, and both agree with the generalization.

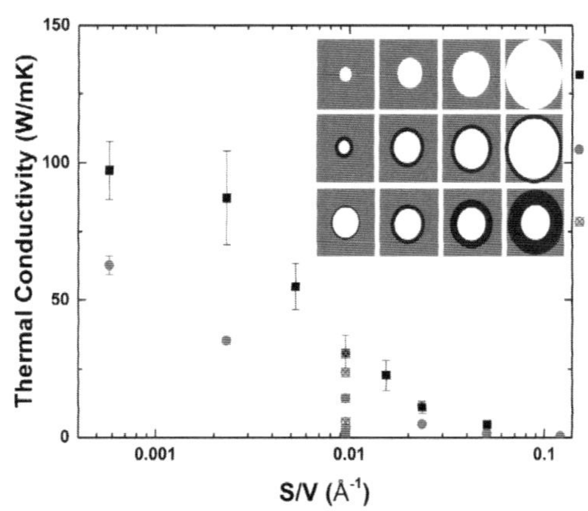

Figure 4. Thermal conductivity of nano-porous Silicon with or without amorphous shell around the pores as a function of the ratio $S/\mathcal{V}$, with S the surface of the pore. With this parameter, results do not merge on the same curve.

6. Conclusion

Nanoporosity can be used to tune the thermal conductivity by controlling phonons scattering on a dense network of interface. This usually dramatically reduces thermal transport. Moreover, the amorphization of the material which can occur around the pores during the elaboration of nanoporous structures contributes to further decrease the thermal conductivity. This effect is more pronounced when the amorphized zone is large. Adding a shell of amorphous Silicon around very small spherical nano-pores in crystalline Silicon, one can even reach sub-amorphous thermal conductivities while keeping a continuous part of crystalline phase. It is expected that a similar effect would be observed with oxidation around the pores (amorphous SiO_2).

Finally, the behavior of the thermal conductivity of nanoporous structures with and without amorphous shells can be universalized with the surface to volume ratio by considering the amorphous/crystalline interface as the main scattering surface for phonons when a shell encloses the pores. This shows that void and amorphous phase globally disturb thermal transport in composite nano-structures in the same way.

Acknowledgments

Authors would like to acknowledge to Dr. A. France-Lanord (CEA Material Design) for support and for many helpful discussions through-out this study. Calculations were performed on the Ermione cluster (IJL-LEMTA).

References

[1] David L. Andrews and Gregory D. Scholes, editors. *Comprehensive nanoscience and technology*. Elsevier, Acad. Press, Amsterdam, 2011.

[2] J.-H. Lee, J. C. Grossman, J. Reed, and G. Galli. Lattice thermal conductivity of nanoporous Si: Molecular dynamics study. *Applied Physics Letters*, 91(22):223110, 2007.

[3] Valentin Jean. Modélisation du transport de phonons dans les semi-conducteurs nanostructurés. 2015.

[4] David Song and Gang Chen. Thermal conductivity of periodic microporous silicon films. *Applied Physics Letters*, 84(5):687, 2004.

[5] M. Isaiev, S. Tutashkonko, V. Jean, K. Termentzidis, T. Nychyporuk, D. Andrusenko, O. Marty, R. M. Burbelo, D. Lacroix, and V. Lysenko. Thermal conductivity of meso-porous germanium. *Applied Physics Letters*, 105(3):031912, July 2014.

[6] A. France-Lanord, S. Merabia, T. Albaret, D. Lacroix, and K. Termentzidis. Thermal properties of amorphous/crystalline silicon superlattices. *Journal of Physics: Condensed Matter*, 26(35):355801, September 2014.

[7] A. France-Lanord, E. Blandre, T. Albaret, S. Merabia, D. Lacroix, and K. Termentzidis. Atomistic amorphous/crystalline interface modelling for superlattices and core/shell nanowires. *Journal of Physics: Condensed Matter*, 26(5):055011, February 2014.

[8] E. Blandre, L. Chaput, S. Merabia, D. Lacroix, and K. Termentzidis. Modeling the reduction of thermal conductivity in core/shell and diameter-modulated silicon nanowires. *Physical Review B*, 91(11), March 2015.

[9] William Humphrey, Andrew Dalke, and Klaus Schulten. VMD – Visual Molecular Dynamics. *Journal of Molecular Graphics*, 14:33–38, 1996.

[10] Steve Plimpton. Fast parallel algorithms for short-range molecular dynamics. *Journal of computational physics*, 117(1):1–19, 1995.

[11] Frank H. Stillinger and Thomas A. Weber. Computer simulation of local order in condensed phases of silicon. *Physical Review B*, 31(8):5262–5271, April 1985.

[12] R.L.C. Vink, G.T. Barkema, W.F. van der Weg, and Normand Mousseau. Fitting the StillingerWeber potential to amorphous silicon. *Journal of Non-Crystalline Solids*, 282(2-3):248–255, April 2001.

[13] M. Verdier, K. Termentzidis, and D. Lacroix. Crystalline-amorphous silicon nano-composites: Nano-pores and nano-inclusions impact on the thermal conductivity. *Journal of Applied Physics*, 119(17):175104, May 2016.

Thermal transport study across interface "nanostructured solid surface / fluid" by photoacoustic technique

K. Voitenko[1a], M. Isaiev[1,2], A. Pastushenko[3], D. Andrusenko[1], A. Kuzmich[1], V. Lysenko[3], and R. Burbelo[1]

[1] Taras Shevchenko National University of Kyiv, 64/13, Volodymyrska Str., Kiev 01601, Ukraine
[2] Laboratoire LEMTA, Faculté des Sciences et Technologies, CNRS UMR 7563, Université de Lorraine, BP 70239, 54506 Vandoeuvre les Nancy Cedex, France
[3] University of Lyon, Nanotechnology Institute of Lyon, UMR CNRS 5270, INSA de Lyon, F69621, France

[a] voitenkokateryna@gmail.com

Abstract. In the paper the experimental study of heat transport across the interface "porous silicon/liquid" by photoacoustic technique is reported. Two cases with and without liquid covering of porous silicon surface were considered. Thermal perturbations were excited at the surface of porous silicon as a result of absorption of the light with modulated intensity. The resulting thermal-elastic stresses arising in the system were registered with piezoelectric transducer. The amplitude-frequency dependencies of the voltage on the piezoelectric electrodes were measured. The presence of the liquid film leads to decreasing of the amplitude of photoacoustic signal as a result of the thermal energy evacuation from the porous silicon into the liquid. The experimental dependencies were fitted with the results of simulation that takes into account heat fluxes separation at the porous silicon/liquid interface. With the presented method one can precisely measure heat fluxes transferred from the solid into contacting fluid. Moreover, the presented approach can be easily adopted for the thermal conductivity study of the different nanofluids as well as thermal resistance at the interface nanostructured solid/fluid.

1. Introduction

Interfacial phenomena play a significant role in chemistry, material science, biology etc. In particular, heat, mass and charge transfers across interfaces are crucial for formation of nanostructures [1], functionalization of surfaces [2], in the medicine [3]–[5], and so on. Therefore, the prediction and tuning of the interface properties are essential for the various applied areas.

Photoacoustic technique is an efficient tool for the investigation of the thermal transport properties of different systems. It can be easily applied for the study of materials with different structures and morphologies [6, 7]. This method is nondestructive and requires no special sample preparation procedure since the thermal field excitation is contactless. Moreover, since the parameters of photoacoustic signal depend only on the portion of absorbed energy, it is efficient for the study of highly scattered media such as nanostructured materials and biological tissues [8-11]. It is well-known that there is huge amount of different kinds of photoacoustic setup configurations. However, they could be classified as direct and non-direct. In non-direct ones the acoustic response is measured in surrounding media. In the direct setups thermoelastic stresses occurring in the sample are detected.

In the paper we discuss the application of photoacoustic phenomena to study the heat transfer across nanostructured solid/liquid interface. Piezoelectric configuration was chosen as a photoacoustic technique with important sensitivity [12, 13]. An experimental configuration where the heat perturbation is excited at the solid/fluid interface was used. As a substrate, the layers of porous silicon with different porosities grown on monocrystalline wafer were chosen as samples with well-defined and well-tuned properties. Since porous silicon surface is hydrophobic, water can be used as a liquid not penetrating in the pores. The experimental measurements of the amplitude-frequency dependencies with/without covering fluid of the nanostructured surface were performed. It allows us to distinguish the heat fluxes passing into solid and, as a result, in the liquid.

2. Experimental details

2.1. Sample preparation

The porous silicon samples were fabricated by electrochemical etching of p+-type monocrystalline Si wafer doped with boron (resistivity about 10-20 mOhm·cm). The etching solution consists the hydrofluoric acid and ethanol (HF (49%):C_2H_5OH=1:1). The initial thickness of silicon wafer was equal to 500 μm. For obtaining porous silicon with different porosity (45%, 55%, and 65%), different anodic current densities were applied (50 mA/cm^2, 100 mA/cm^2, 200 mA/cm^2, respectively). The time of etching was set to perform the thickness of porous silicon layer to be equal to 50 μm. The morphologies of resulting porous silicon layers were observed via SEM, in particular SEM image of the prepared porous silicon sample with porosity 65% is shown in figure 1b.

2.2. Experimental setup

The photoacoustic signal formation in the porous silicon with/without water film on the sample surface was studied in configuration presented in figure 1a. The thickness of water cover layer was equal to 1 mm. The porous silicon surface was irradiated by square wave modulated light of LED with wavelength λ = 470 nm. As a result of light absorption the thermal distribution was induced in the sample, thus the thermoelastic stresses were created. The resulting deformations of the system were registered with piezoelectric transducer.

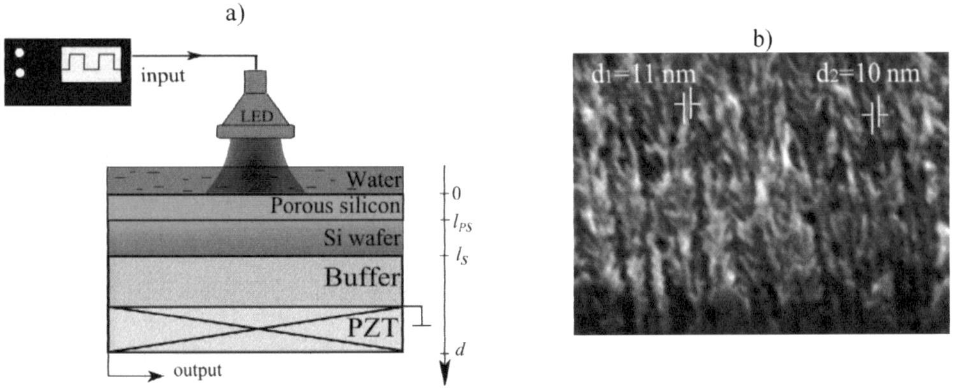

Figure 1. Schematic sketch-view of the geometry of the studied system (a). The SEM image of the prepared porous silicon layer with porosity 65% (b).

The dielectric buffer layer with low thermal conductivity and thermal expansion coefficient values was used to avoid penetration of heat fluxes in piezoelectric transducer (PZT). The thickness of the buffer layer was chosen to be equal to 1 mm to perform the conditions that the total thickness of the layered structure is much bigger than the thickness of the sample. Amplitude and phase shift of photoacoustic responses (voltage on piezoelectric electrodes) were experimentally measured in the frequency range from 10 Hz to 3000 Hz.

3. Results and discussions

The typical amplitude-frequency dependencies for the cases with/without covering liquid are presented in figure 2. The solid lines present the results of simulations based on the evaluation of the temperature distribution in a multilayered structure with planar periodic heat source located at the interface porous silicon/liquid [14]. In the case of low frequencies (less than 100 Hz), the thermal wavelengths are bigger than the thickness of porous silicon layer and almost all heat energy is transferred into the crystalline silicon. As one can see from the figure, the amplitudes of the signals for the cases with and without water layer are approximately the same for the frequencies <100 Hz. This fact confirms equivalence of the optical conditions for the both cases.

For the treatment of experimental data, we used an approximation of quasi-stationary deformations of the system. In frames of this approximation, voltage difference on the piezoelectric electrodes could be written in the following form [14]

$$\Delta\varphi \sim \int_0^{l_s} \frac{\alpha_T E}{1-\nu}\theta(s)g(s)ds \qquad (1)$$

here α_T is the thermal expansion coefficient of the material, E is Young's modulus, ν is the Poisson coefficient, $\theta(s)$ is the temperature distribution in the structure, $g(s)$ is the function which describes the efficiency of the voltage generation depending on the point s where the elastic source was applied, l_s is the thickness of Si wafer. The spatial distribution of the variable component of temperature $\theta(s)$ in the layered structure was evaluated by solving the heat conduction equation with the corresponding boundary conditions [15]. The expression (1) was used to fit experimental amplitude–frequency dependencies. Thus, thermal conductivities χ_{PSi} of porous silicon samples with different porosity were evaluated (0,50 W/m·K, 0,33 W/m·K, 0,25 W/m·K for porosity 45%, 55%, 65%).

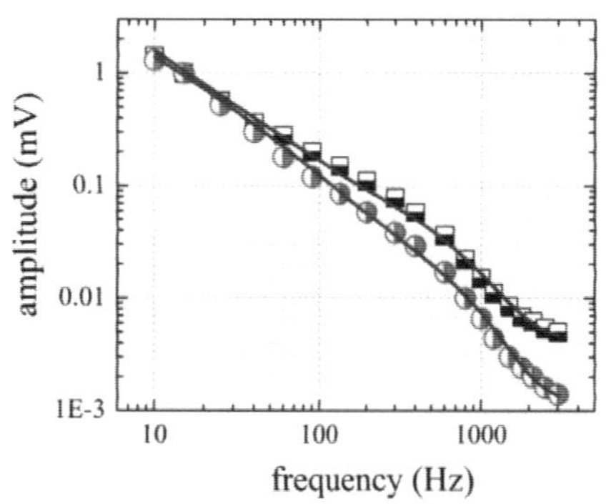

Figure 2. The amplitude–frequency dependencies of the photoacoustic signal measured for porous silicon sample (porosity 55%) for the cases with (circles)/without (squares) presence of the water layer. The results of simulation are presented with continuous lines.

Additionally, the comparison of these the cases with/without water layer shows that in the region of high frequencies (larger than 100 Hz) the amplitude of the signal is bigger for the case of absence of water film. In the case of the presence of water film, the part of the thermal energy is evacuated from the porous silicon into the water.

The dependencies of spatial distribution of ac-temperature component amplitude for the cases with/without water layer for excitation frequencies 100 Hz (a) and 2000 Hz (b) are presented in the figure 3. The evaluation of ac-temperature component was made with the use of Fourier law based heat conduction equation with a planar periodic heat source located at the interface porous silicon/fluid. The system of coordinate was chosen the same as in Fig 1a: origin is located at the interface and abscissa is directed in the depth of porous silicon. As one can see from the figure, for the case of excitation frequency equal to 100 Hz, the important temperature rise appears in the crystalline silicon. For the higher modulation frequency, the temperature rise in the sample is negligible.

a) b)

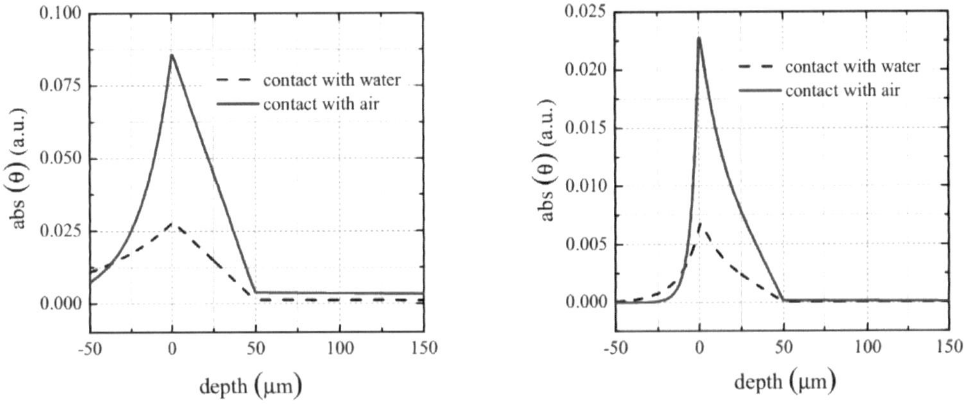

Figure 3. The spatial distribution of the amplitude of ac-temperature component for different frequencies: 100 Hz (a), 2000 Hz (b).

The deformations of the considered system caused by light and amplitude of photoacoustic signal are proportional to the thermal energy in the porous silicon and silicon substrate. Therefore, one can conclude that from the ratio (ε) of amplitudes (figure 4a) of photoacoustic signal with and without water film the portion of the thermal energy transferred in the water from the porous silicon can be define. In particularly, ratio of energy transferred in the water to the energy transferred in the porous silicon (γ) can be define as follows (fig. 4b):

$$\gamma = (1-\varepsilon)\, {e_w}/{e_{PSi}}, \qquad (2)$$

where $e_w = \sqrt{c_w \rho_w \chi_w}$, $e_{PSi} = \sqrt{c_{PSi} \rho_{PSi} \chi_{PSi}}$ are thermal effusivities of water and porous silicon respectively, c_w, ρ_w χ_w, c_{PSi}, ρ_{PSi}, χ_{PSi} are specific heat capacities, densities and thermal conductivities of water and porous silicon, respectively.

a) b)

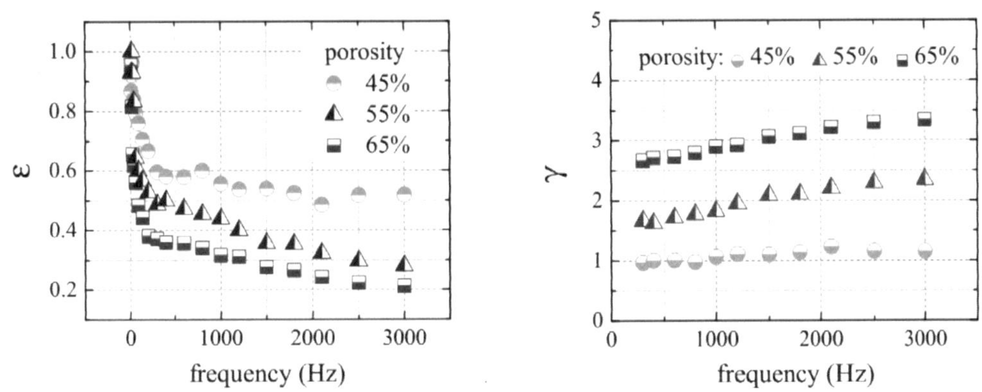

Figure 4. The ratio of photoacoustic signal amplitudes for the cases with/without water film as a function of modulation frequency for different porosities of porous silicon layer (a). Frequency dependencies of γ coefficient for samples with different porosities (b).

As one can see from the figure 4, the presence of water film covering the surface leads to increasing of the efficiency of heat transport from the porous silicon into the water at the high frequencies. Moreover, this tendency is more pronounced for porous silicon layers with a high value of porosity due to their lower thermal conductivity.

4. Conclusions

Photoacoustic technique with piezoelectric registration was adopted for the study of heat transport across the interface nanostructured solid/fluid. As a solid substrate porous silicon layer with well-define and well-tuned properties was chosen. The amplitude-frequency dependencies of the photoacoustic signal were measured for two cases with and without water layer covering solid surface. It was shown that the presence of the liquid layer leads to the decreasing of the amplitude of photoacoustic signal amplitude. Correlation of the experimental results with the result of simulation demonstrates that this decreasing arises because of thermal energy evacuation from the solid into substrate. The heat energy evacuated from the porous silicon increased with porosity, and for the case of the porosity equals to 65 % the portion of thermal energy transferred in the water is more than in 3 times bigger the portion transferred in porous silicon. The presented approach can be easily adopted for thermal properties study of different fluids and for thermal resistance of the interfaces nanostructured solid/liquid.

Acknowledgment

Research work described in this paper was partially performed in frames of Marie Skłodowska-Curie Research and Innovation Staff Exchange CARTHER (proposal # 690945). Authors acknowledge European Commission for financial support.

References

[1] Garah M El, Perone R C, Bonilla A S, Haar S, Campitiello M, Gutierrez R, Cuniberti G, Masiero S, Ciesielski A, and Samorì P 2015 *Chem. Commun.* **51** 11677.
[2] Goicochea J V, Hu M, Michel B, and Poulikakos D, 2011 *J. Heat Transfer* **133** 082401
[3] Fragneto-Cusani G 2001 *J. Phys. Condens. Matter* **13** 4973
[4] Ge Z, Cahill D G, and Braun P V 2006 *Phys. Rev. Lett.* **96** 18
[5] Dna S P, Van Grinsven B, Vanden Bon N, Strauven H, Grieten L, Murib M, Janssens S D, Haenen K, Scho M J, Jime K L, Vermeeren V, Ameloot M, Michiels L, Thoelen R, De Ceuninck W, and Wagner P 2012 *ACS Nano* **6** 2712
[6] Isaiev M, Tutashkonko S, Jean V, Termentzidis K, Nychyporuk T, Andrusenko D, Marty O, Burbelo R M, Lacroix D, Lysenko V 2014 *Appl. Phys. Lett.* **105** 031912
[7] Tytarenko A I, Andrusenko D A, Kuzmich A G, Gavril I V, Skryshevskii V A, Isaiev M V, Burbelo R M 2014 *Tech. Phys. Lett.* **40** 188
[8] Zhong J, Yang S, Zheng X, Zhou T, Xing D 2013 *Nanomedicine (Lond)* **8** 903
[9] Ohl C, Gao F, Feng X, Zheng Y 2014 *J. Biomed. Opt.* **19** 067006
[10] Anees P, Joseph J, Sreejith S, Menon N V, Kang Y, Yu S W-K, Ajayaghosh A, Zhao Y 2016 *Chem. Sci. Royal Society of Chemistry* **7** 4110
[11] Subochev P, Fiks I, Frenz M, Turchin I 2016 *Laser Phys. Lett. IOP Publishing* **13** 25605
[12] Alekseev S, Andrusenko D, Burbelo R, Isaiev M, Kuzmich A 2011 *J. Phys. Conf. Ser.* **278** 012003
[13] Isaiev M, Voitenko K, Doroshchuk V, Andrusenko D, Kuzmich A, Skryshevskii V, Lysenko V, Burbelo R. 2015 *Phys. Procedia* **70** 586
[14] Isaiev M V, Andrusenko D A, Tytarenko A I, Kuzmich A G, Lysenko V V, and Burbelo R M, 2014 *Int. J. Thermophys.* **35** 2341
[15] Andrusenko D, Isaiev M, Kuzmich A, Lysenko V, Burbelo R 2012 *Nanoscale Res. Lett.* **7** 411

IOP Conf. Series: Journal of Physics: Conf. Series **785** (2017) 012011 doi:10.1088/1742-6596/785/1/012011

On the dependence of the thermal conductivity of width-modulated nanowires on the number of modulations

Xanthippi Zianni[1]*, Konstantinos Termentzidis[2] and David Lacroix[2]

[1] Dept. of Aircraft Technology, Technological Educational Institution of Sterea Ellada, 34400 Psachna, Greece

[2] Université de Lorraine, LEMTA, CNRS, UMR 7563, Faculté des Sciences et Technologies, BP 70239, 54506 Vandoeuvre les Nancy cedex, FRANCE

Electronic address: * xzianni@teiste.gr

Abstract. Our previous Monte Carlo simulations on the thermal conductivity of width-modulated nanowires indicated two distinct dependences of the decrease of the thermal conductivity κ relative to that of the non-modulated nanowire: (i) in the case of multiple constrictions κ scales with the nanowire transmissivity, (ii) in the case of a single constriction κ is determined by the ballistic constriction resistance. Here, we report on the transition between the two regimes. We discuss the thermal conductivity of width modulated nanowires as a function of the number of modulations. Phenomenology has been derived to interpret the MC simulations.

1. Introduction

Controlling the heat transfer is highly required in order to increase the efficiency of many applications in thermoelectrics, optoelectronics, nanoelectronics etc. Understanding the thermal conductivity of nanostructures is therefore a major research issue of scientific and technological interest. In nanostructures, nanoscale non-uniformity may cause phonon modification and phonon scattering that drastically affect the thermal conductivity. It is important to understand the dependence of the thermal conductivity on the underlying mechanisms. Our work aims to contribute to this need.

Porous materials and low-dimensional structures with characteristic dimensions above the quantum confinement regime, show reduced thermal conductivity due to enhanced phonon boundary scattering [1-5]. Width-modulated nanowires have been previously proposed as efficient thermoelectric materials [6-7]. It has been theoretically predicted that the transport properties of electrons and phonons are modified resulting in enhanced thermal to electrical energy conversion efficiency. In the quantum confinement regime, the phonon transmission coefficient and phonon energy states are modified so that the thermal conductivity decreases [8-10]. Above the quantum confinement regime, the thermal conductivity of modulated nanowires was studied within the kinetic theory for phonons using an approximate model for boundary scattering [11]. A big reduction of the thermal conductivity was found due to enhanced boundary scattering and the decrease of the phonon transmissivity by the nanowire width-modulation. Phonon Monte Carlo (MC) simulations confirmed this evidence [12]. In addition they provided further insight and more quantitative description because this method is free of assumptions on the boundary scattering [12,13].

Our previous MC simulations on the thermal conductivity of width-modulated nanowires [12] indicated two distinct dependences of the decrease of the thermal conductivity κ relative to that of the non-modulated nanowire: (i) in the case of multiple constrictions κ scales with the phonon transmissivity, (ii) in the case of a single constriction κ is determined by the ballistic constriction resistance. Here, we report on the transition between the two regimes. We discuss the thermal conductivity of width modulated nanowires as a function of the number of modulations. Phenomenology has been derived to interpret the MC simulations.

2. Methodology and results

We have performed phonon MC simulations for the phonon transport using a previously developed simulation tools. The numerical method is extensively described in [12,14,15]. It lies on the resolution of the Boltzmann transport equation in the frame of the relaxation time approximation. Silicon dispersion properties have been used for the sampling of phonon frequency and group velocity, as well as for the calculation of the scattering relaxation times. The characteristic relaxation time is given according to the 'Matthiessen' summation rule. It takes into account: normal (N), umklapp (U) and impurity (I) scattering rates according to Holland [16] formalism. Here, we present simulations at room temperature. Diffusive boundary scattering has been assumed. At room temperature, the phonon's wavelength is about 1 nm, depending on the way it is evaluated. Thus, the specularity parameter ($p = \exp\,(-16\pi^3\delta^2/\lambda^2)$, where δ is the roughness parameter and λ is the wavelength) is close to 0, meaning that every single reflection is diffuse.

The thermal conductivity of a nanowire with width modulated by a constriction is lower than that of a non-modulated nanowire [12]. This is to be expected since the constriction introduces extra boundary scattering. The thinner is the constriction the bigger is the decrease of the thermal conductivity. This dependence was found for all nanowire lengths as shown in Figure 1a, where the thermal conductivity is plotted as a function of the nanowire length for different constriction widths.

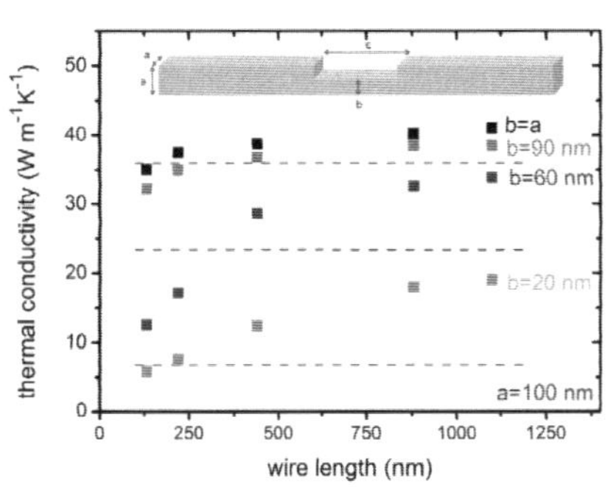

Figure 1. The thermal conductivity of a nanowire modulated by a single constriction versus the nanowire length for various values of the constriction width b (squares). The length of the constriction is c=10 nm. Periodically modulated nanowires with the same parameters a and b

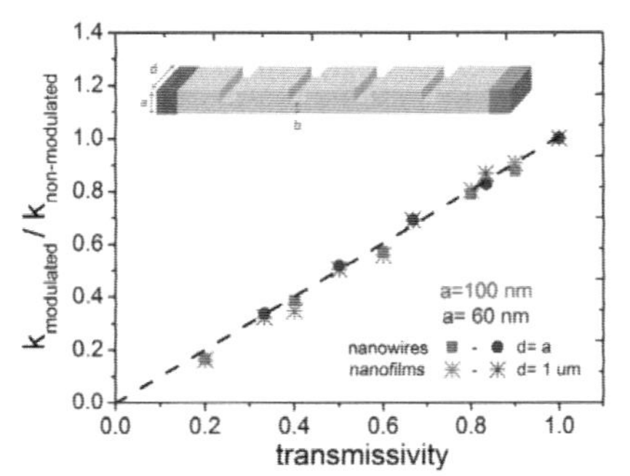

Figure 2. The thermal conductivity of two periodically modulated nanowires with a = d, and a = 100 nm (red squares) and a=60 nm (blue dots) versus the transmissivity. The length of the constriction is c=10 nm.

(dashed lines).

In Figure 1 it can be seen that the length dependence of the thermal conductivity is similar to that of the non-modulated nanowire for not very thin constrictions. It becomes though significantly steeper when a constriction becomes thin. Based on the steepness of the length dependence of the thermal conductivity, a transition between two regimes can be distinguished. This transition is clearly shown in the temperature profile across the modulated nanowire [12]. In non-very thin constrictions, e.g. 90 nm, the temperature profile is continuous and phonon transport is diffusive. When the constriction becomes thinner, e.g. 60 nm, a small discontinuity appears. This is the transition point. In thinner constrictions, e.g. 20 nm, the temperature profile is discontinuous. The discontinuity in the temperature profile is the signature of ballistic phonon transport across the constriction. A critical width can be identified below which phonon transport becomes ballistic. The ballistic transport is quantitatively described by a constriction thermal conductance K_C [12]. This ballistic conductance is analogous to the thermal contact conductance. It is determined by the width mismatch between the nanowire and the constriction, namely the constriction transmissivity $Tr= A_c /A_w$ where A_w and A_c respectively denote the cross sections of the wire and of the constriction. For the geometries of Figure 1, it holds: $Tr= b/a$, where a and b are respectively the widths of the nanowire and of the constriction. The constriction thermal conductance K_C scales with the transmissivity Tr according the relation: $K_C \propto (1-Tr)^{-1}$ [12].

The analysis of the data showed that in this regime the effective thermal conductivity of the modulated nanowire can be decomposed into two contributions: (i) of the thermal conductivity of the non-modulated segments of the nanowire, and (ii) of the contribution of the constriction thermal conductance. The data can be interpreted by the approximate relation: $R^{th}_{mw} = R^{th}_{C} + R^{th}_{W}$, where R^{th}_{mw}, R^{th}_{C}, R^{th}_{W} are the thermal resistances of the modulated nanowire, of the constriction and of the non-modulated nanowires respectively. It holds that $K_C=1/R_C$.

In the presence of multiple constrictions the thermal conductivity decreases more significantly than by a single constriction. For adequate modulation density, the decrease is proportional to the nanowire transmissivity [12]. This behavior is illustrated in Figure 2. In this regime, the thermal conductivity cannot be decomposed to distinct contributions and the signature of the single constriction thermal resistance is absent. The temperature profile across the nanowire has been found continuous. It can be thereby concluded that phonon transport is diffusive.

Let us now explore the transition from the ballistic constriction resistance behavior to the transmissivity behavior that takes place as the number of constrictions increases. We have performed simulations in nanowires with fixed length and with variable number of constrictions, i.e. variable modulation length as shown in the schematic representation in Figure 3. In Figure 4a it is shown the thermal conductivity of a modulated nanowire of fixed length versus the number of constrictions modulating its width. The different curves correspond to different constriction widths.

Figure 3. Schematic representation of a nanowire partially modulated by constrictions.

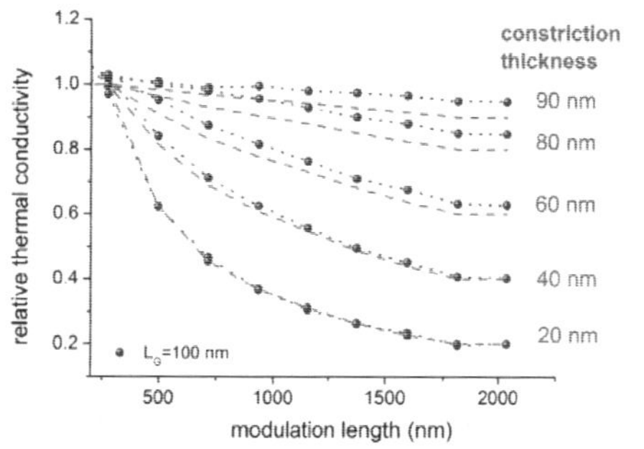

Figure 4a The thermal conductivity of the modulated nanowire relative to that of the non-modulated nanowire versus the modulation length for various constriction thicknesses. The red lines are the fittings of the phenomenological equation.

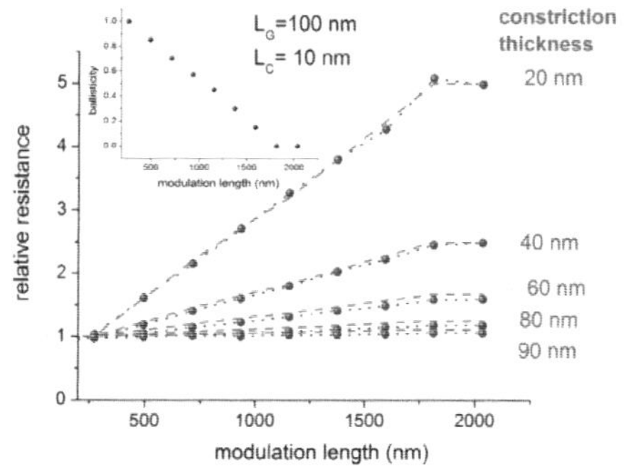

Figure 4b As in Figure 3a, for the relative resistance. The inset shows the fitting parameter χ versus the modulation length.

In Figure 4a, it can be seen that the thermal conductivity decreases gradually with a decreasing rate as the number of constrictions increases. It decreases more as more constrictions are added. This is because more constrictions provide more scatterers for phonons. In Figure 4b it is shown the resistance as a function of the number of constrictions. It can be seen that the resistance increases first linearly with the number of constrictions. The linear dependence implies that each time a constriction is added, a resistance proportional to R_c is added to the overall resistance. This holds until a certain resistance plateau is reached. At this plateau the thermal conductivity is proportional to the nanowire transmissivity. In this regime the thermal conductivity is independent of the number of constrictions. This behavior can be interpreted by the phenomenological equation:

$$R = R_w(L)\left[1+(1-\chi)(Tr^{-1}-1)\right]+\chi N R_C$$

where $R_w(L)$ is the thermal resistance of the non-modulated segment of the nanowire with length L, N is the number of constriction and χ is the ballisticity of the nanowire. The parameter χ is a fitting parameter that is determined by the Monte Carlo simulations. It is plotted in the inset of Figure 4b. The data of this fitting plot were determined by the curve of Figure 4a for constriction thickness 20 nm. It has been found to successfully interpret all the curves shown in Figure 4. Parameter χ decreases nearly linearly up to the critical plateau. For N=1 it holds χ=1 which corresponds to the single constriction ballistic regime. For $N>N_{plateau}$, χ~0 which corresponds to the scaling law with the transmissivity.

The calculated thermal conductivity and thermal resistance using the above phenomenological equation is shown in Figure 4 with red lines. It can be seen that the thermal resistance is interpreted extremely good by the above phenomenology. The thermal conductivity is also interpreted equally well for thin constrictions below the transition regime. Deviations have been found for not very thin constrictions. This is to be expected since above the critical thickness, the constriction thermal resistance R_C that is used in the phenomenological equation is not a well defined [nano]. The above equation indicates that when a small number of constrictions are present in the nanowire, they act as ballistic thermal resistances. As the number of constrictions increases, they gradually contribute more to the nanowire boundary scattering and to reduce the conduction of phonons though the modulated nanowire. For adequate number of constrictions the geometrical transmissivty of the structure determines the thermal conductivity.

3. Main conclusion

The thermal conductivity of width-modulated nanowires decreases gradually with increasing number of modulations. The decrease can be interpreted by the additional constriction thermal resistance with increasing number of constrictions. After a certain number of constrictions, the thermal conductivity depends on the transmissivity of the structure and not on the number of constrictions.

References

[1] Heron J S, Fournier T, Mingo N and Bourgeois O 2010 Mesoscopic size effects on the thermal conductance of silicon nanowire *Nano Letters* 9 (2009) 1861.
[2] Heremans J P, Dresselhaus M S, Bell L E and Morelli D T 2013 Reflections on thermoelectrics *Nature Nanotechnology* 8 (2013) 471
[3] Jean V, Fumeron S, Termentzidis K, Tutashkonko S and Lacroix D 2014 Monte Carlo simulations of phonon transport in nanoporous silicon and germanium *Journal of Applied Physics* 115 (2014) 024304
[4] Jain A, Yu Y-J, McGaughey J H 2013 Phonon transport in periodic silicon nanoporous films with feature sizes greater than 100 nm *Physical Review B* 87 (2013) 195301
[5] Liu L-C and Huang M-J 2011 *International Journal of Thermal Sciences* (2011) 15479
[6] Zianni X 2010 Diameter-modulated nanowires as candidates for high thermoelectric energy conversion efficiency *Appl. Phys. Lett.* 97 (2010) 233106
[7] Zianni X 2011 Efficient thermoelectric energy conversion on quasi-localized electron states in diameter modulated nanowires *Nanoscale Research Letters* 6 (2011) 286
[8] Zianni X 2012 The effect of the modulation shape in the ballistic thermal conductance of modulated nanowires *Journal of Solid State Chemistry* 193 (2012) 53
[9] Nika D L, Cocemasov A I, Isacova C I, Balandin A A, Fomin V M, and Schmidt O G 2012 *Physical Review B* 85 (2012) 205439
[10] Termentzidis K, Barreteau T, Ni Y, Merabia S, Zianni X, Chalopin Y, Chantrenne P and Volz S 2013 Modulated SiC nanowires : Molecular Dynamics study of their thermal properties *Phys. Rev. B* 87 (2013) 125410
[11] Zianni X and Chantrenne P 2013 Thermal conductivity of diameter-modulated nanowires within a frequency-dependent model for the phonon boundary scattering *Journal of Electronic Materials* 42 (2013) 1509
[12] Zianni X, Jean V, Termentzidis K and Lacroix D 2014 Scaling behavior of the thermal conductivity of width-modulated nanowires and nanofilms for heat transfer control at the nanoscale *Nanotechnology* 25 (2014) 465402
[13] Jean V, Fumeron S, Termentzidis K, Zianni X, Lacroix D 2015 Monte Carlo simulations of phonon transport in Si nanowires with constrictions *International Journal of Heat and Mass Transfer* 86 (2015) 648
[14] Lacroix D, Joulain K, Lemonnier D 2005 Monte Carlo transient phonon transport in silicon and germanium at nanoscales *Physical Review B* 72 (2005) 064305
[15] Peraud J-P M, Hadjiconstantinou N G 2011 Efficient simulation of multidimensional phonon transport using energy-based variance-reduced Monte Carlo formulations *Physical Review B* 84 (2011) 205331
[16] Holland M G 1963 Analysis of Lattice Thermal Conductivity *Physical Review* 132 (1963) 2461

Institute of Physics
Dirac House, Temple Back
Bristol BS1 6BE UK

CURRAN ASSOCIATES INC.
proceedings.com

ISSN: 1742-6588
ISBN 978-1-5108-3534-4

9 781510 835344

3rd International Workshop on Metamaterial-By-Design 2017

Journal of Physics: Conference Series Volume 963

Madrid, Spain
14 - 15 December 2017

IOP
Institute of Physics